Diagnostic Testing in Advanced Chemistry

Diagnostic Testing in Advanced Chemistry

A. BROOKES Ph.D.,

Head of Chemistry, Withernsea High School, Withernsea, North Humberside.

W. A. H. SCOTT Ph.D.,

Lecturer in Science Education, The University of Bath.

HODDER AND STOUGHTON

LONDON SYDNEY AUCKLAND TORONTO

Brookes, A
  Diagnostic testing in advanced chemistry.
  Test volume
  1. Chemistry—Examinations, questions, etc.
  I. Title   II. Scott, W A H
  540'.76   QD42

  ISBN 0-340-24190-X

  Complete Vol.
  ISBN 0-340-24189-6

First Printed 1979

Printed in Great Britain for Hodder and Stoughton Educational,
a division of Hodder and Stoughton Ltd,
Mill Road, Dunton Green, Sevenoaks, Kent, by
J. W. Arrowsmith Ltd., Bristol BS3 2NT.

Contents

Acknowledgements

We should like to express our thanks to all those people who have contributed to this book in however small a way. To Mr. Tony Hull of the Fitzmaurice Grammar School, Bradford-on-Avon. however, we owe an especial debt of gratitude for his comments, advice and constructive suggestions. Our thanks too must go to the students in the schools which helped us. We would like to stress that any errors or areas of doubt which remain in the book are our responsibility alone. We would welcome comments on such matters.

Preface

The test items and diagnostic notes in this book may be used in a variety of ways. They can provide information for teachers about different parts of the course; they can be used to provoke class discussion or they may be used by students as personal checks on their knowledge, abilities and progress. It is possible to use them for revision purposes as parts of the course are reviewed.

Each item is associated with a set of diagnostic notes, which bears a title describing the intellectual skill which is likely to be used in answering the item. Care is needed, however, in the interpretation of these descriptions. The skill which has to be exercised by a student in order to answer an item depends upon the background of the student, and on his knowledge and experience. A comprehension exercise for one person might be simple recall for another. In our view, the 'knowledge' items reflect what it is reasonable to expect an advanced student to know.

The three categories which are used in this book are based on Bloom's Taxonomy of Educational Objectives. The categories are:

Knowledge —of facts, theories, principles, laws, generalizations, experimental procedures and techniques.

Comprehension—of theories and laws, experimental procedures, techniques and design; the interpretation of information in familiar situations.

Application —of knowledge and principles to novel and unfamiliar situations, of numerical, non-numerical and graphical data; the interpretation and evaluation of experimental results and the making of predictions.

Notes on units and nomenclature

The physicochemical quantities used in this book, with one exception, are expressed in coherent SI units or other units accepted for general use. The exception is the quantity pressure, which is expressed in mmHg or atm, depending on the circumstances. Only one commonly accepted name and symbol has been used for each quantity except in the area of thermodynamics, where the expressions 'enthalpy change of combustion', etc. and 'heat of combustion', etc. are both employed in the text.

The organic chemical nomenclature employed in the book is that recommended in the A.S.E. publication of 1972, *Chemical Nomenclature, Symbols and Terminology*. In respect of inorganic nomenclature, however, the procedure which has been adopted by many of the examining boards at advanced level has been followed in that, although A.S.E. recommended names are used, the following exceptions are made:

1. The oxidation state of an element has been omitted where the element has only one common oxidation state;
2. only common names have been used for oxoacids and oxoanions, e.g. sulphuric acid and sulphate.

With all the nomenclature, however, where we feel that confusion might arise alternative names and/or formulae are given.

I Atomic Structure

1 How many neutrons are there in one atom of the isotope $_n^m X$?

    A m
    B m−n
    C n
    D n−m

2 How many electrons are there in the neutral atom of the isotope $_n^m X$?

    A m
    B m−n
    C n
    D n−m

3 Which one of the following ions contains d electrons? (Atomic numbers: K = 19, Ca = 20, Sc = 21, Fe = 26)

    A K^+
    B Ca^{2+}
    C Sc^{3+}
    D Fe^{2+}

4 Where in the Periodic Table would you place an element whose electronic configuration is $1s^2\, 2s^2\, 2p^6\, 3s^2\, 3p^6\, 4s^2\, 3d^5$?

    A Group II
    B Group V
    C Group VII
    D none of the above

5 Which one of the following isotopes contains the same number of protons and neutrons?

    A $_{18}^{40}Ar$
    B $_{19}^{40}K$
    C $_{20}^{40}Ca$
    D $_{21}^{44}Sc$

6 Which one of the following is *not* deflected in either a magnetic or an electric field?

    A alpha particle
    B beta particle
    C gamma ray
    D proton

7 Which one of the following has the greatest charge/mass ratio?

    A   proton
    B   neutron
    C   electron
    D   alpha particle

8 Which one of the following is produced if $^{218}_{84}$Po loses an alpha particle, a beta particle and a gamma ray?

    A $^{214}_{81}$Tl
    B $^{214}_{83}$Bi
    C $^{213}_{82}$Pb
    D $^{215}_{82}$Pb

9 What is the species X, formed in this reaction

$$^{14}_{7}N + {}^{4}_{2}He \rightarrow {}^{17}_{8}O + X?$$

    A   proton
    B   gamma ray
    C   neutron
    D   electron

10 Which one of the following combinations of alpha and beta particles emitted from $^{228}_{90}$Th would produce $^{212}_{83}$Bi?

    A $4\alpha + 1\beta$
    B $1\alpha + 4\beta$
    C $4\alpha + 7\beta$
    D $8\alpha + 1\beta$

11 Which one of these statements, concerning radioactivity is *untrue*?

    A   Alpha particles are helium nuclei.
    B   Gamma rays travel at the velocity of light in a vacuum.
    C   Beta particles are high energy valency electrons.
    D   Gamma rays are emitted by a nucleus in an excited state.

12 The intensity of radiation emitted by an element falls to one eighth of its original value in 48 days. What is the half-life of the element?

    A   6 days
    B   12 days
    C   16 days
    D   24 days

13 Frequencies in the ultraviolet region of the hydrogen spectrum can be calculated from the expression:

$$\text{frequency} = Rc\left(\frac{1}{n_1^2} - \frac{1}{n_2^2}\right)$$

where R is a constant, c is the speed of light, n_1 is the quantum number of the energy level to which the electron moves and n_2 is the quantum number of the energy level from which the electron moves. Which one of the following transitions produces radiation of the largest frequency?

A $(n_2 = 2) \rightarrow (n_1 = 1)$
B $(n_2 = 3) \rightarrow (n_1 = 1)$
C $(n_2 = 3) \rightarrow (n_1 = 2)$
D $(n_2 = 5) \rightarrow (n_1 = 2)$

14 Which one of the following transitions corresponds to the second line of the first series in the hydrogen spectrum?

A $(n = 2) \rightarrow (n = 1)$
B $(n = 3) \rightarrow (n = 1)$
C $(n = 4) \rightarrow (n = 2)$
D $(n = 5) \rightarrow (n = 3)$

15 Which one of the following statements concerning this reaction is *incorrect*?

$$^{235}_{92}U + ^1_0n \rightarrow ^{95}_{42}Mo + ^{139}_{57}La + 2^1_0n + 7_{-1}^0e$$

A This is a nuclear fission reaction.
B A large amount of energy is taken in during this change.
C The reaction rate could be controlled by absorbing neutrons.
D The neutrons produced can propagate a chain reaction.

16 The Co^{2+} ion has the electronic configuration (Argon) $3d^7$. How many unpaired electrons might it possess?

A 7
B 5
C 3
D 2

17 Paramagnetism is a phenomenon involving the attraction of a chemical species into a magnetic field. The phenomenon is generally caused by the presence of unpaired electrons. Paramagnetism will be exhibited by

A sodium.
B magnesium.
C argon.
D zinc.

3

II Structure and Bonding

1 Which one of the following is a correct statement about ionic lattices?

    A   The co-ordination numbers of sodium and chlorine in the sodium chloride lattice are both eight.

    B   The sodium chloride lattice is composed of two inter-penetrating face-centred cubic lattices.

    C   The co-ordination numbers of caesium and chlorine in the caesium chloride lattice are both six.

    D   Caesium chloride has a face-centred cubic lattice (cubic close-packed).

2 X-rays of wavelength λ are incident to the surface of a crystal at an angle of θ. The distance between layers in the crystal is d and a wave reflected from the second layer is two wavelengths out of phase with a ray reflected from the upper layer. Which one of the following is the correct expression for the wavelength of the X-rays?

    A $\lambda = \dfrac{d}{2}\sin\theta$

    B $\lambda = 2d\sin\theta$

    C $\lambda = d\sin\theta$

    D $\lambda = 4d\sin\theta$

3 Which one of the following will *not* conduct an electric current?

    A   graphite

    B   solid calcium fluoride

    C   molten sodium chloride

    D   aqueous potassium iodide solution

4 Which one of the following substances will have the smallest degree of ionic character?

    A $NaCl$

    B $MgCl_2$

    C $RbCl$

    D $AlCl_3$

5 Which one of the following molecules contains at least one bond angle which is *less* than the tetrahedral angle (109°28′)?

    A CO_2

    B BF_3

    C SiH_4

    D PF_5

6 Which one of the following species has *no* lone pairs of electrons on the underlined atom?

A H$_2$$\underline{S}$
B $\underline{N}$F$_3$
C $\underline{B}$F$_3$
D H$_3$$\underline{O}$$^+$

7 Which one of the following sets of data concerning the shape of molecules is *incorrect*?

	No. of lone pairs on central atom	No. of bonding pairs	Shape
A	0	3	trigonal bipyramidal
B	1	3	trigonal pyramidal
C	0	6	octahedral
D	2	2	planar, V-shaped

8 Which one of the following compounds does *not* have an overall dipole moment?

A HF
B H$_2$O
C NH$_3$
D CCl$_4$

9 In which one of these substances will hydrogen bonding be present to the largest extent?

A H$_2$
B H$_2$O
C HI
D H$_2$Se

10 In which one of these compounds are the constituent ions isoelectronic?

A LiF
B Na$_2$O
C SrI$_2$
D CaCl$_2$

11 Which one of the following species contains two co-ordinate linkages per molecule/ion?

A Al$_2$Cl$_6$
B NH$_4$Cl
C C$_2$H$_5$OH
D [Fe(H$_2$O)$_6$]$^{3+}$

12 Which one of the following statements concerning water in different physical states is correct?

    A   Hydrogen bonding between water molecules cannot occur above 100 °C, 1 atm. pressure.

    B   In ice, each oxygen atom is surrounded by four hydrogen atoms.

    C   The hydrogen–oxygen covalent bond in ice is longer than a hydrogen–oxygen bond formed by hydrogen bonding.

    D   In ice, molecules are able to slide easily over each other.

13 Which one of the following liquids would be unaffected if a stream of it passed by an electrostatically charged glass rod?

    A   water

    B   hexane

    C   propanone

    D   trichloromethane

14 The extent of scattering of X-rays by a crystal is due to

    A   the size of the unit cell.

    B   the number of electrons possessed by the atoms in the crystal.

    C   the shape of the atoms.

    D   the number of atoms in the unit cell.

15 Which one of the following statements is true of a close-packed structure?

    A   The co-ordination number of each atom is a function of atomic number.

    B   Only one crystal system is classed as close-packed.

    C   Alkali metals have close-packed structures.

    D   In close-packed structures empty space within the crystal is minimized.

16 Which one of the following substances would you expect to crystallize in an isotropic form?

    A   NaBr

    B $NaNO_3$

    C $CaCO_3$

    D CaC_2

17 Which one of the following molecules contains *at least* one bond angle of 180°?

    A BF_3

    B H_2O

    C $BeCl_2$

    D HNO_3

18 The following substances all contain a nitrogen–nitrogen bond: N_2, N_2H_4, $N_2(CH_3)_2$. Which one of the following is written in order of increasing N–N bond length?

A $N_2(CH_3)_2$, N_2H_4, N_2
B N_2, $N_2(CH_3)_2$, N_2H_4
C N_2H_4, $N_2(CH_3)_2$, N_2
D N_2, N_2H_4, $N_2(CH_3)_2$

Questions 19 to 25 refer to the incomplete dot and cross diagram of the amino acid

$$HS-CH_2-CH-COOH$$
$$|$$
$$NH_2$$

Some of the electrons in the bonds are represented by dots and crosses whilst others are represented by a line and a letter (p, q, r, s and t). The questions concern the number of electrons constituting the bonds, their nominal origin and the lone pairs within the molecule.

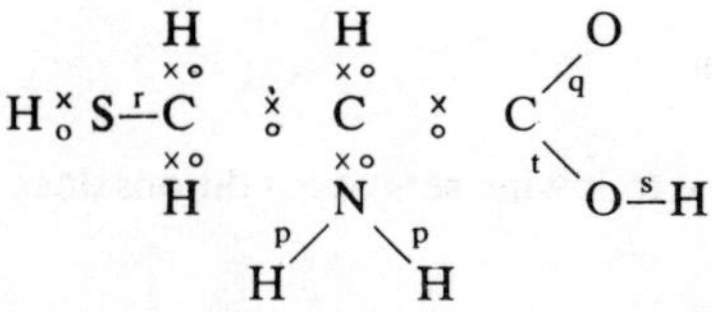

Use the responses A to D to answer questions 19 to 25. Each response may be used once, more than once or not at all.

A   1
B   2
C   3
D   4

How many electrons are there in each of these bonds?

19 p

20 q

21 r

How many electrons are regarded as coming from the oxygen atom into each of the following bonds?

22 s

23 q

In the molecule shown, how many lone pairs of electrons are there on each of these atoms?

24 sulphur

25 nitrogen

7

III States of Matter

1 0.29 g of propanone ($M_r = 58$) occupies a fixed volume in a gas syringe at 140 °C. 0.23 g of a volatile liquid Y will also occupy the same volume in the syringe at the same temperature and pressure. What is the relative molecular mass of Y?

A $\dfrac{0.23 \times 58}{0.29}$

B $\dfrac{0.29 \times 58}{0.23}$

C $\dfrac{0.23}{0.29 \times 58}$

D $0.23 \times 58 \times 0.29$

2 Under which one of the following sets of conditions does a gas deviate most from ideal behaviour?

    A low pressure, low temperature
    B high pressure, low temperature
    C high pressure, high temperature
    D low pressure, high temperature

3 Hydrogen and oxygen are allowed to diffuse through the same hole. Which one of the following is the correct value for the ratio *rate of diffusion of hydrogen : rate of diffusion of oxygen*?

    A $1:16$
    B $1:4$
    C $1:1$
    D $4:1$

4 A fixed mass of gas at temperature T(K) and pressure p (mmHg) occupies a volume V (dm³). What is the volume of this fixed mass of gas in dm³ if it is measured at s.t.p.?

A $\dfrac{V \times T}{273 \times p \times 760}$

B $\dfrac{273 \times V}{760 \times T \times p}$

C $\dfrac{273 \times V \times p}{760 \times T}$

D $\dfrac{760 \times V \times p}{273 \times T}$

5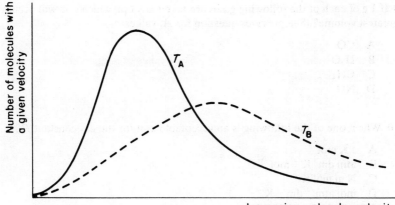

Increasing molecular velocity

Which one of the following is a true statement about these curves representing the distribution of molecular velocities in a sample of gas?

A   Raising the temperature lowers the most probable molecular velocity.
B   Temperature T_A is greater than temperature T_B.
C   Raising the temperature reduces the number of molecules with the most probable velocity.
D   There is a wider distribution of velocities at a lower temperature.

6 Which one of the following changes will increase the vapour pressure in a sealed vessel containing water?

A   increasing the quantity of water
B   adding salt to the water
C   doubling the volume of the vessel
D   raising the temperature

7 One mole of N_2O_4 is placed in a sealed vessel and heated to $X\,°C$. At this temperature it is 50% dissociated into NO_2. What is the ratio of partial pressures $N_2O_4 : NO_2$ if the dissociation is represented by the equation

$$N_2O_4(g) \rightleftharpoons 2NO_2(g)?$$

A   1:2
B   1:1
C   2:1
D   4:1

8 Which one of the following has the largest mean square velocity $(\overline{c^2})$ at 25 °C? (A_r values: H = 1, C = 12, N = 14, O = 16, S = 32)

A $O_2(g)$
B $NH_3(g)$
C $H_2S(g)$
D $CO(g)$

9

9 If 1 g of each of the following gases are taken at s.t.p., which one will occupy the greatest volume? (See previous question for A_r values)

    A CO
    B H_2O
    C CH_4
    D NO

10 Which one of the following is an acceptable unit for the gas constant R?

    A $J\,K\,mol^{-1}$
    B $atm\,dm^3\,K^{-1}\,mol^{-1}$
    C $N\,m\,mol\,K^{-1}$
    D $mol\,atm^{-1}\,dm^{-3}\,K^{-1}$

11 Which one of the following is a valid alternative to the ideal gas equation: $pV = nRT$?

Where M = relative molecular mass
       d = density
       m = mass of gas

    A $T = \dfrac{pMV}{mR}$

    B $T = \dfrac{pd}{RM}$

    C $T = \dfrac{pm}{Rd}$

    D $T = \dfrac{pmV}{MR}$

12 Which one of the following reactions, all of which go to completion, results in a decrease in volume of one third, if the volumes of reactants and products are measured at the same temperature and pressure?

    A $2\,A_2(g) + B_2(g) \rightarrow 2\,A_2B(g)$
    B $A_2(g) + B_2(g) \rightarrow A_2B_2(g)$
    C $A_2(g) + 2\,B_2(g) \rightarrow A_2B_4(g)$
    D $2\,A_2(g) + 5\,B_2(g) \rightarrow 2\,A_2B_5(g)$

13 At s.t.p. which one of the following would occupy approximately $22.4\ dm^3$? (M_r of $O_2(g) = 32$)

    A 16 g of $O_2(g)$
    B 0.33 mole of $O_3(g)$
    C 16 g of $O(g)$
    D 0.5 mole of $O(g)$

14 Which one of the following conditions *must* apply when determining relative molecular masses of volatile liquids by the injection of a small volume of liquid into a heated gas syringe?

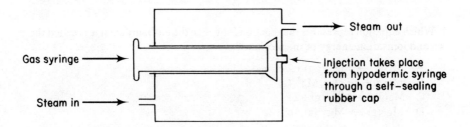

A   The liquid's boiling point should be above the temperature of the syringe.
B   The gas syringe should be completely empty at the start of the experiment.
C   The liquid should have a low heat of vaporization.
D   The substance under test should be stable at the temperature of the gas syringe.

IV Periodicity

1 Which equation represents the reaction for which the enthalpy change is called the 'second ionisation energy of magnesium'?

    A $Mg(s) \rightarrow Mg^{2+}(aq) + 2\,e^-$
    B $Mg(g) + 2\,e^- \rightarrow Mg^{2-}(g)$
    C $Mg(g) \rightarrow Mg^{2+}(g) + 2\,e^-$
    D $Mg^+(g) \rightarrow Mg^{2+}(g) + e^-$

2 Which one of the following decreases in value when descending Group I?

    A atomic radius
    B ionic radius (monopositive ion)
    C first ionization energy
    D electropositivity

3 When a graph of heat of fusion for elements is plotted against atomic number a regular pattern of peaks and troughs is seen. Which set of atomic numbers is those of the elements for which the peaks occur?

    A 6, 14, 32
    B 3, 11, 19
    C 9, 17, 35
    D 5, 13, 31

4 The electronic configuration of the first element in the first transition series is (Argon) $4s^2\,3d^1$. The electronic configuration of chromium, the fourth member of the series will be

    A (Argon) $4s^2\,3d^1\,4p^3$
    B (Argon) $4s^1\,3d^5$
    C (Argon) $4s^2\,3d^4$
    D (Argon) $4s^1\,3d^2\,4p^3$

5 When descending a group of elements in the Periodic Table, the radius of the atom

    A increases.
    B decreases.
    C remains approximately the same.
    D increases and subsequently decreases.

6 Which of these oxides is most acidic?

    A P_4O_{10}
    B Cl_2O_7
    C SiO_2
    D Al_2O_3

7 Which of the following best explains the fact that the 'head element' of a group is not typical of that group?

    A   The element has a greater melting point than those below it.
    B   The element has a lower first ionisation energy than others in the group.
    C   The element has the smallest and most electronegative atom in the group.
    D   The element has a low atomic number.

8 Which one of these sets of elements is written in order of increasing atomic radius?

    A   N O F
    B   Br Cl F
    C   O F Cl
    D   F O S

9 Which one of these statements concerning radii of atoms and ions is *incorrect*?

    A $Cl^- > Cl$
    B $Si^{4-} > Si$
    C $H^- > H^+$
    D $Na^+ > Na$

10 'Group IV elements of the Periodic Table (carbon, silicon, germanium, tin and lead) become progressively more metallic in character as the group is descended'. Which one of the following does *not* support the above statement?

    A   Carbon and silicon, in the diamond form, show negligible conductivity; germanium is a semi-conductor, and tin and lead exhibit high conductivities.
    B   Carbon is attacked by oxidizing acids, silicon by alkalis, germanium by oxidizing acids and alkalis, tin by hot alkali and dilute acid. Lead is even more reactive than tin.
    C   Allotropes of the elements at the top of the group exist, but the elements at the bottom do not exhibit allotropic forms.
    D   The formation of ions by simple electron loss is unknown at the top of the group, but is the norm at the bottom.

11 Three successive elements (increasing atomic number) have these values for their first ionisation energies: 1680, 2080, 494 (kJ mol^{-1}). Which one of the following sets represents the three elements?

    A   O F Ne
    B   F Ne Na
    C   Ne Na Mg
    D   Na Mg Al

12 Which one of these pairs of elements exhibits a 'diagonal relationship'?

    A   Cl and Kr
    B   K and Ca
    C   Li and Mg
    D   Na and Be

13 Tellurium is an element in Group VI of the Periodic Table, with a relative atomic mass greater than that of sulphur. Which one of these statements about tellurium is *incorrect*?

    A It has an atomic radius greater than that of oxygen.
    B There are two s electrons and four p electrons in its outermost energy level.
    C Its first ionisation energy is greater than that of sulphur.
    D It will form an oxide of formula TeO_2.

14 The following species are isoelectronic (contain the same number of electrons). The correct order of their increasing radii is

    A K^+, Ar, Ca^{2+}.
    B Ar, K^+, Ca^{2+}.
    C Ca^{2+}, Ar, K^+.
    D Ca^{2+}, K^+, Ar.

15 A suggested definition of a transition metal is ... 'an element which has an incomplete energy level other than the outer one'. In the case of the first transition series, the 'incomplete level' is the third and the 'outer one' the fourth. Using this definition, which one of the following is *not* a transition metal? (Atomic No. K = 19, Sc = 21, Mn = 25, Zn = 30)

    A Manganese
    B Potassium
    C Scandium
    D Zinc

16 A better definition of a transition metal is 'an element which gives rise to compounds containing an incomplete 'd' sub-level of electrons'. Using this definition, which one of the following can be regarded as a transition metal ion? (Atomic No. Sc = 21, V = 23, Zn = 30)

    A Sc^+
    B Zn^{2+}
    C Sc^{3+}
    D V^{3+}

17 Which one of the following pairs of species is isoelectronic?

    A Mg^{2+} and NaH
    B CN^- and N_2
    C N_2 and Li_3N
    D LiF and MgH_2

18 Which of the following sets of ionisation energies (kJ mol^{-1}) best fits the pattern associated with a Group IV element?

	1st I.E.	2nd I.E.	3rd I.E.	4th I.E.	5th I.E.
A	1260	2300	3800	5200	6500
B	420	3100	4400	5900	8000
C	580	1800	2700	11600	14800
D	760	1500	3300	4400	9000

19 Which one of the following *must* contain an odd number of electrons?

    A   compounds of Group I elements
    B   binary (non-ionic) compounds of adjacent elements in the same period
    C   ions possessing one negative charge
    D   species containing an odd number of atoms

Use the Periodic Table shown here to answer questions 20 to 22. The letters on the table do not represent the symbols of any element. You may use the letters A to H once, more than once or not at all.

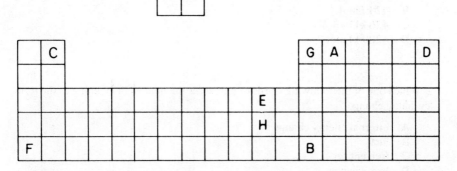

20 Which element would give rise to two common oxidation states with values +I and +III?

21 Which element would give the most acidic oxide?

22 Which element's properties most resemble those of aluminium?

V Thermodynamics

1 0.1 mole of an acid is dissolved in water to make $100\,cm^3$ of solution. The temperature rises by $10\,°C$. What is the enthalpy change for the process, assuming the mass of the solution to be $100\,g$ and the specific heat capacity of the solution to be $4.2\,J\,K^{-1}\,g^{-1}$?

    A $4.2\,kJ\,mol^{-1}$
    B $42\,kJ\,mol^{-1}$
    C $420\,kJ\,mol^{-1}$
    D $4200\,kJ\,mol^{-1}$

2 Which one of the following words or phrases is the nearest in meaning to the term 'enthalpy change'?

    A   internal energy change
    B   heat
    C   free energy
    D   heat change

3 For which one of the following reactions in the gas phase is the enthalpy change numerically equal in value to the O—H bond energy?
    A $H_2 + \frac{1}{2}O_2 \rightarrow H_2O$
    B $H_2 + O_2 \rightarrow H_2O_2$
    C $O^{2-} + H^+ \rightarrow OH^-$
    D $O + H \rightarrow OH$

4 Which one of the following would you expect to have the largest bond energy?

    A $C\equiv C$
    B $C=C$
    C $C-C$
    D $C=O$

5 Which one of the following is a necessary condition for the measurement of the standard enthalpy change of combustion of an element?

    A   The atmospheric pressure should be 760 mmHg.
    B   The element should be pure.
    C   The temperature should be 298 K.
    D   The products should be gaseous.

6 Which one of the following is an exothermic process?

    A   the first ionization energy of sodium
    B   the second electron affinity of oxygen
    C   the sum of the first and second ionization energies of magnesium
    D   the first ionization energy of the $Cl^{2-}(g)$ ion

Questions 7 and 8. The following diagram represents a Born–Haber cycle for the formation of MX(s) and MX(aq) from the elements M, a Group I metal, and X_2, a halogen.

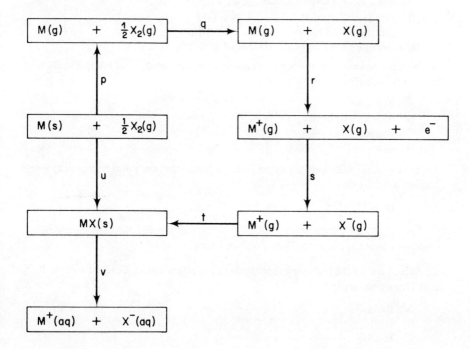

7 In which one of the following *pairs* of steps *must both* be exothermic?

    A   p+t
    B   q+r
    C   s+p
    D   s+t

8 In which one of the following *pairs* of steps *might both* steps be exothermic?

    A   p+s
    B   s+r
    C   u+v
    D   v+q

9 In the Hess's law diagram shown below, $\Delta H_1^\ominus$ is known. Which one of the other sets of information shown is necessary for the calculation of $\Delta H_2^\ominus$?

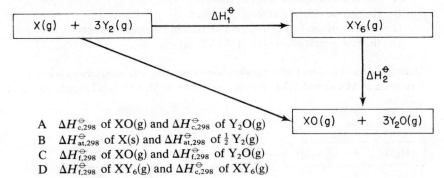

A $\Delta H_{c,298}^\ominus$ of XO(g) and $\Delta H_{c,298}^\ominus$ of Y_2O(g)
B $\Delta H_{at,298}^\ominus$ of X(s) and $\Delta H_{at,298}^\ominus$ of $\frac{1}{2}$ Y_2(g)
C $\Delta H_{f,298}^\ominus$ of XO(g) and $\Delta H_{f,298}^\ominus$ of Y_2O(g)
D $\Delta H_{f,298}^\ominus$ of XY_6(g) and $\Delta H_{c,298}^\ominus$ of XY_6(g)

10 Which one of the following is unstable both kinetically and energetically with respect to its oxidation product(s)?

A kerosine
B white phosphorus
C diamond
D sugar

11 Which of the following elements is stable both kinetically and energetically with respect to its oxide(s)?

A gold
B iron
C sodium
D hydrogen

12 Which one of the following represents the substance in its standard state at 10 °C and 1 atm pressure?

A Br_2(g)
B H(g)
C H_2O(l)
D CO_2(s)

13 The standard heats of combustion of graphite and carbon monoxide are −393 and −283 kJ mol^{-1} respectively. What is the standard heat of formation of carbon monoxide in kJ mol^{-1}?

A 173
B −110
C −676
D −1069

14 In which of the following processes will the enthalpy change be exothermic?

A Na(s) $\rightarrow$ Na(g)
B Ca(s) $\rightarrow$ Ca^+(g)
C Br(g)+e^- $\rightarrow$ Br^-(g)
D I_2(s) $\rightarrow$ I_2(g)

15 When 25 cm^3 of 1 M NaOH is neutralized with an equal volume of 1 M HCl the temperature of the mixture rises by 6.8 °C. What will be the temperature change if 50 cm^3 of 0.5 M NaOH is neutralized with an equal volume of 0.5 M HCl? (Assume heat losses to be negligible in each case.)

    A   The answer cannot be determined from the information given.
    B   3.4 °C
    C   6.8 °C
    D   13.6 °C

16 In an experiment to find the heat of combustion of ethanol ($M_r = 46$), 0.23 g were completely burned. The heat produced was totally absorbed in raising the temperature of 100 g of water by 10 °C. What is the heat of combustion of ethanol in J mol^{-1} if the specific heat capacity of water is 4.18 J K^{-1} g^{-1}?

    A $\dfrac{1000 \times 10 \times 4.18}{0.23}$

    B $\dfrac{100 \times 10 \times 4.18 \times 46}{0.23}$

    C $\dfrac{100 \times 10 \times 4.18 \times 0.23}{46}$

    D $\dfrac{100 \times 10 \times 4.18}{46}$

17 Which one of the following quantities is *not* required in the calculation of the enthalpy change of formation of sodium chloride using a Born–Haber cycle?

    A   lattice energy of sodium chloride
    B   first ionization energy of sodium
    C   electron affinity of chlorine
    D   hydration energy of sodium chloride

18 Which equation represents the reaction for which the enthalpy change is called the standard heat of formation of calcium fluoride?

    A $Ca(g) + F_2(g) \rightarrow CaF_2(g)$
    B $Ca(s) + F_2(g) \rightarrow CaF_2(s)$
    C $Ca^{2+}(g) + 2\,F^-(g) \rightarrow CaF_2(s)$
    D $Ca^{2+}(s) + 2\,F^-(g) \rightarrow CaF_2(s)$

19 The heat of combustion of propane, $C_3H_8(g)$, is −2220 kJ mol^{-1}. Which one of the following quantities of propane ($M_r = 44$) will evolve 555 kJ on complete combustion?

    A   14.66 g
    B   22.4 dm^3 measured at s.t.p.
    C   22 g
    D   5.6 dm^3 measured at s.t.p.

VI Equilibria

1 Which are the correct units for expressing the equilibrium constant, K_c, for the reaction:

$$RS_2(g) \rightleftharpoons 2\,S(g) + R(g)?$$

A $mol^{-2}\,dm^{-6}$
B $mol^2\,dm^{-6}$
C $mol^{-1}\,dm^3$
D $mol^3\,dm^{-9}$

2 For the exothermic forward reaction:

$$N_2(g) + 3\,H_2(g) \rightleftharpoons 2\,NH_3(g)$$

Which one of the following will increase the concentration of ammonia at equilibrium?

A   decreasing the concentration of
B   increasing the pressure
C   using a positive catalyst
D   raising the temperature

3 In which one of the following reactions will an increase in total pressure cause an increase in the concentration of the underlined species?

A $H_2(g) + Cl_2(g) \rightleftharpoons 2\,\underline{HCl}(g)$
B $2\,SO_2(g) + O_2(g) \rightleftharpoons 2\,\underline{SO_3}(g)$
C $CaCO_3(s) \rightleftharpoons CaO(s) + \underline{CO_2}(g)$
D $N_2O_4(g) \rightleftharpoons 2\,\underline{NO_2}(g)$

4 In a reaction, carried out in $1\,dm^3$ of solution,

$$A + B \rightleftharpoons C + D$$

there were initially a moles of A and b moles of B present only. At equilibrium, there were x moles of C present. What is the equilibrium constant, K_c, for the reaction?

A $\dfrac{x^2}{ab}$

B $\dfrac{x}{(a-x)(b-x)}$

C $\dfrac{x^2}{(a-x)(b-x)}$

D $\dfrac{ab}{x}$

5 What are the units of K_p for the reaction:

$$N_2(g) + 3\,H_2(g) \rightleftharpoons 2\,NH_3(g)$$

if pressures are measured in atm?

A atm^2
B atm^{-2}
C atm^{-1}
D no units

6 In the reaction:

$$Y(g) \rightleftharpoons 2Z(g)\ \Delta H \text{ (forward reaction) is exothermic}$$

which one of the following increases the equilibrium constant?

A using a catalyst
B increasing the pressure
C raising the temperature
D lowering the temperature

7 A quantity of $N_2O_4(g)$ was placed in a sealed flask and heated to 25 °C. At equilibrium, the pressure in the vessel was 3 atm. The vessel contained a mixture of 2 mole of $NO_2(g)$ and 1 mole of $N_2O_4(g)$. The equation for the dissociation is:

$$N_2O_4(g) \rightleftharpoons 2\,NO_2(g)$$

What is the value of K_p in atm for the reaction?

A 0.25
B 2
C 4
D 9

8 A quantity of ammonium chloride was heated in a sealed vessel to 200 °C and it dissociated according to the equation:

$$NH_4Cl(s) \rightleftharpoons NH_3(g) + HCl(g).$$

What is the total pressure in atm in the vessel at equilibrium if the value of K_p at 200 °C is x atm^2?

A $\sqrt{x}$
B $2\sqrt{x}$
C x
D x^2

9 Which one of the following expressions represents the solubility product of lead(II) chloride, $PbCl_2$?

A $[Pb^{2+}(aq)][Cl^-(aq)]\ mol^2\ dm^{-6}$
B $[Pb^{2+}(aq)][Cl^-(aq)]^2\ mol^3\ dm^{-9}$
C $[Pb^{2+}(aq)]^{1/2}[Cl^-(aq)]\ mol^{3/2}\ dm^{-9/2}$
D $2[Pb^{2+}(aq)][Cl^-(aq)]\ mol^2\ dm^{-6}$

10 For the reaction:

$$Ag^+(aq) + Fe^{2+}(aq) \rightleftharpoons Ag(s) + Fe^{3+}(aq),$$

K_c will be equal to

A $\dfrac{[Fe^{3+}(aq)][Ag(s)]}{[Ag^+(aq)][Fe^{2+}(aq)]}$

B $\dfrac{[Ag^+(aq)][Fe^{3+}(aq)]}{[Fe^{2+}(aq)]}$ mol dm^{-3}

C $\dfrac{[Ag(s)][Fe^{3+}(aq)]^3}{[Ag^+(aq)][Fe^{2+}(aq)]^2}$ mol dm^{-3}

D $\dfrac{[Fe^{3+}(aq)]}{[Ag^+(aq)][Fe^{2+}(aq)]}$ mol^{-1} dm^3

11 Which one of the following is a way of expressing the concentration of a gas; where R is the gas constant, V is the volume of the gas and n is the number of moles of gas?

A $\dfrac{V}{n}$

B $\dfrac{R}{n}$

C $\dfrac{n}{R}$

D $\dfrac{n}{V}$

12 In the equilibrium:

$$2\,NO_2(g) \rightleftharpoons 2\,NO(g) + O_2(g)$$

which one of the following will move the position of equilibrium to the right?

A the addition of oxygen
B the addition of nitrogen monoxide
C a decrease in pressure
D an increase in pressure

13 Which one of the following is a correct statement about the equilibrium:

$$CaCO_3(s) \rightleftharpoons CaO(s) + CO_2(g)?$$

A $[CaCO_3(s)] = 0$
B $[CaCO_3(s)] = [CaO(s)]$
C $K_c = [CO_2(g)]$
D $[CaO(s)] = 0$

14 Hydrogen and oxygen are mixed in the proportion 2 : 1 without reaction occurring. Each gas was initially at 750 mmHg pressure and no pressure change or volume change occurs on mixing. What is the partial pressure of oxygen in mmHg after mixing?

A   250
B   500
C   750
D   1000

VII Kinetics

Questions 1 to 7 concern the reaction of three substances W, X and Y. The information in the table below refers to the initial concentrations of the reactants and the time taken for the reaction to go to completion. Use this information to answer questions 1 to 7.

Reaction	Concentration of W/mol dm^{-3}	Concentration of X/mol dm^{-3}	Concentration of Y/mol dm^{-3}	Time taken for the reaction to go to completion/s
a	0.4	0.24	0.01	152 ± 6
b	0.8	0.24	0.01	73 ± 4
c	1.2	0.24	0.01	52 ± 5
d	0.4	0.48	0.01	148 ± 8
e	0.4	0.12	0.01	154 ± 8
f	0.4	0.24	0.02	78 ± 3
g	0.4	0.24	0.03	48 ± 4

1 The rate of reaction is proportional to the initial concentration of W raised to the power

    A 0.
    B $\frac{1}{2}$.
    C 1.
    D 2.

2 As the initial concentration of X is increased, the rate of reaction

    A increases in proportion to [X].
    B increases in proportion to $[X]^2$.
    C decreases in proportion to [X].
    D stays the same.

3 The order of reaction with respect to Y is

    A zero.
    B first.
    C second.
    D fractional.

4 The overall order for the reaction is

    A zero.
    B first.
    C second.
    D third.

5 The overall rate expression for the reaction is: Rate equals

    A $k[X]^1[Y]^1$ mol dm^{-3} s^{-1}.
    B $k[X]^1[W]^1$ mol dm^{-3} s^{-1}.
    C $k[Y]^1[W]^1$ mol dm^{-3} s^{-1}.
    D $k[W]^3[Y]^{-1}$ mol dm^{-3} s^{-1}.

6 In the rate expression, the units of k are

    A mol dm^{-3} s^{-1}.
    B mol^{-1} dm^3 s^{-1}.
    C mol dm^{-3} s.
    D mol^{-1} dm^3 s.

7 What name is given to k in the rate expression?

    A rate order
    B rate concentration
    C rate molecularity
    D rate constant

8 How would you best follow the course of the reaction:

$$2\,HI(g) \rightleftharpoons H_2(g) + I_2(g)?$$

    A conductivity measurements
    B pH measurements
    C colorimetry
    D a sealed gas syringe/flask unit

9 By referring to the energy diagram shown below, decide which one of the following statements is true.

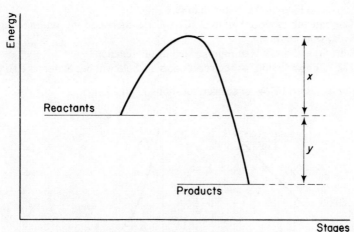

    A $(x+y)$ kJ mol^{-1} is the enthalpy change for the reaction.
    B The reaction described is endothermic.
    C In such energy diagrams, if $x > y$, the reaction must be endothermic.
    D x kJ mol^{-1} is the activation energy of the reaction.

10 The reaction:

$$A(g) \rightarrow B(g)$$

occurs via a certain route (i). The rate along this route is slow. The addition of a catalyst increases the rate substantially because the catalyst

    A provides a new route for the reaction to occur along where the activation energy is less than for route (i).
    B lowers the standard enthalpy change for the reaction.
    C reduces the activation energy along route (i).
    D causes the activation energy along route (i) to become negative in value.

11 What effect does a catalyst have on the standard enthalpy change for a reaction?

    A increases it
    B reduces it
    C changes it so as to make it equal to the activation energy
    D leaves it unchanged

12 Bromoalkanes can be hydrolysed by two routes (a) and (b):

    (a) step i: $R-Br \rightarrow R^+ + Br^-$
          step ii: $R^+ + H_2O \rightarrow ROH + H^+$
    (b) $R-Br + H_2O \rightarrow ROH + H^+ + Br^-$

In route (b), a transition stage is involved where the alkyl chain R has both the bromine atom and the water molecule attached to it. The approach of the water molecule and the loss of the bromine atom, as the ion, occur in the same step whereas in route (a) there are two distinct stages in the reaction.

  Which one of the following statements about route (a) can be deduced from the information given above?

    A Step (i) is likely to be faster than step (ii).
    B The rate of reaction in step (ii) will be increased by adding sodium hydroxide solution.
    C Step (ii) is the rate determining step of the reaction.
    D The rate expression for the reaction as a whole will be: Rate = $k[RBr]$.

Use the graphs shown below to answer questions 13 to 15.

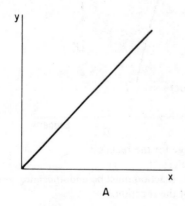

A

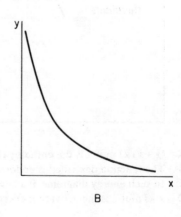

B

26

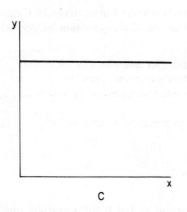

C

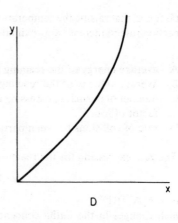

D

13 In which graph could y represent the initial rate of reaction and x represent the initial concentrations of reactant R in a series of reactions which are all first order with respect to R?

14 In which graph does y represent the initial rate of reaction and x the initial concentrations of reactant R in a series of reactions which are zero order with respect to R?

15 In which graph does y represent the concentration of reactant R and x represent the time elapsed during the reaction? The reaction is first order with respect to R.

16 A correct statement about the reaction

$$2\,X + 3\,Y_2 \rightarrow \text{products}$$

is that the

    A   overall order of the reaction is 5.
    B   overall molecularity of the reaction is 3.
    C   rate expression cannot be predicted from the information given.
    D   order of the reaction with respect to X is 2.

17 A positive catalyst
    A   is always chemically changed after a reaction.
    B   can lower the activation energy of a reaction.
    C   is always physically changed after a reaction.
    D   does not change the route for the reaction.

18 A mixture of hydrogen and chlorine gases will react violently if exposed to bright sunlight. The first step in the reaction is

    A $H_2(g) \rightarrow H^+(g) + H^-(g)$.
    B $H_2(g) \rightarrow 2\,H^{\cdot}(g)$.
    C $Cl_2(g) \rightarrow 2\,Cl^{\cdot}(g)$.
    D $Cl_2(g) + e^- \rightarrow Cl_2^-(g)$.

19 It is found that raising the temperature at which a reaction occurs by 20 K causes the reaction rate to increase approximately fourfold. The explanation for this is that the

A   average energy of the reacting particles increases by a factor of four.
B   average velocity of the reacting particles increases fourfold.
C   number of particles possessing a definite minimum energy is increased by a factor of four.
D   rate of collisions between particles increases by a factor of four.

20   The rate expression for the reaction

$$A + B \rightarrow C$$

is Rate $= k[A]^2[B]^{0.5}$.
   Which changes in the initial concentrations of A and B will cause the rate of reaction to increase by a factor of eight?

A $[A] \times 2$; $[B] \times 2$
B $[A] \times 2$; $[B] \times 4$
C $[A]$ constant; $[B] \times 4$
D $[A] \times 4$; $[B]$ constant

21 The following graph was obtained for the decomposition of $N_2O_5(g)$ at 300 °C.

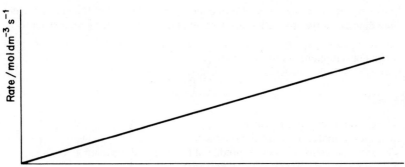

A correct statement about this graph is that

A   it describes a reaction which is second order with respect to $N_2O_5(g)$.
B   the gradient is equal to k, the rate constant.
C   the gradient would not change if the data were collected at 350 °C.
D   the units for the gradient are $mol^{-1} dm^3 s^{-1}$.

28

VIII Colligative Properties

1 Which one of the following is *not* a colligative property of a solution?

    A vapour pressure
    B boiling point
    C freezing point
    D pH

2 Which one of the following is a correct explanation of the cause of the lowering of vapour pressure of a solvent by addition of a non-volatile solute?

    A The boiling point of the solvent is raised.
    B The solute evaporates exerting its own vapour pressure, which is less than that of the solvent.
    C The number of solvent molecules leaving the surface of the solution is reduced by the presence of solute molecules at the surface.
    D The number of solvent molecules returning to the liquid surface is increased by the presence of solute molecules.

3 Which one of the following is a correct statement of Raoult's law concerning the change in vapour pressure of a solvent when a solute is added?
(p_0 = vapour pressure of pure solvent, p_1 = vapour pressure of solution, N = number of moles of solvent, n = number of moles of solute)

    A $\dfrac{p_1 - p_0}{p_0} = \dfrac{n}{N + n}$

    B $p_0 - p_1 = \dfrac{n}{N + n}$

    C $\dfrac{p_0 - p_1}{p_0 + p_1} = \dfrac{n}{N}$

    D $\dfrac{p_0 - p_1}{p_0} = \dfrac{n}{N + n}$

4 The relative lowering of the vapour pressure of a solvent when making a *very* dilute solution by adding a solute is 0.05. If there are 4 moles of solvent, how many moles of solute are added?

    A 0.2
    B 1.0
    C 2.0
    D 5.0

5 What is the exact mole fraction of a solution containing 2 g of a polymer $(M_r = 4000)$ in 37 g of solvent $(M_r = 148)$?

A $\dfrac{\frac{37}{148}}{\frac{2}{4000}}$

B $\dfrac{\frac{2}{4000}}{\frac{37}{148}}$

C $\dfrac{\frac{2}{4000}}{\frac{37}{148} + \frac{2}{4000}}$

D $\dfrac{\frac{37}{148} + \frac{2}{4000}}{\frac{2}{4000}}$

6 Addition of 3 g of a solute X $(M_r = 60)$ to 100 cm^3 of solvent Z causes the same elevation of boiling point as the addition of 6 g of solute Y to 200 cm^3 of solvent Z. What is the relative molecular mass of Y?

A 30
B 60
C 120
D 240

7 If 0.01 mole of each of the following are separately added to 100 g portions of water and the mixtures are stirred in order to dissolve as much of the solid as possible, which mixture will have the highest boiling point?

A glucose
B sodium chloride
C copper(II) oxide
D magnesium chloride

8 Which one of the following does *not* influence the vapour pressure of a dilute solution of a non-volatile solute?

A the temperature of the solution
B the melting point of the solute
C the mole fraction of the solute
D the degree of dissociation of the solute

9 The ebullioscopic constant for water is 0.52 °C kg^{-1}. Which one of these statements concerning solutions at 1 atm pressure is correct?

A 1 mole of sodium chloride (NaCl) in 1 kg of water boils at 100.52 °C.
B 1 mole of sucrose molecules in 1 kg of water boils at 99.48 °C.
C 0.5 mole of sucrose molecules in 1 kg of water boils at 101.04 °C.
D 0.5 mole of glucose molecules in 1 kg of water boils at 100.26 °C.

10 The ebullioscopic constant for benzene is $2.7\,°C\,kg^{-1}$. A solution of 3 g of solute X in 100 g of benzene boils 0.54 °C above the normal boiling point of benzene. What is the relative molecular mass of X?

    A   15000
    B   1500
    C   150
    D   15

11 13 g of a solute X ($M_r = 65$) cause an elevation of boiling point of 0.37 °C if dissolved in 200 cm³ of a solvent P. What mass of solute Y ($M_r = 124$) in g must be dissolved in 400 cm³ of solvent P to cause the same elevation in boiling point, assuming X and Y are undissociated in solution?

    A   12.4
    B   13.0
    C   24.8
    D   49.6

12

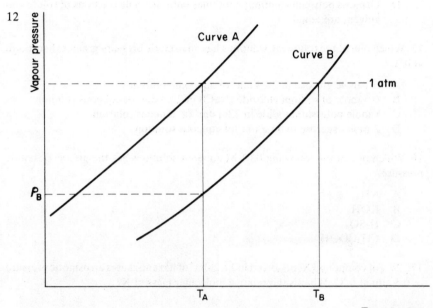

The two curves show data found experimentally when investigating the relationship between temperature and vapour pressure for a pure solvent and a dilute solution of a solute in that solvent.

Which statement concerning the curves is *incorrect*?

    A   Curve A represents pure solvent.
    B   The elevation of boiling point $= T_B - T_A$.
    C   The relative lowering of vapour pressure $= 1 - p_B$.
    D p_B is the vapour pressure of the solution at the boiling point of the pure solvent.

13 Experiments were carried out to calculate the relative molecular masses of solutes using depression of freezing point measurements. Which one of the following solutions would give the correct relative molecular mass for the solute?

A ethanoic acid in hexane
B potassium chloride in water
C phenol in ethoxyethane
D ethanoic acid in water

14 Which one of the following statements concerning osmosis between two solutions is correct?

A Osmosis involves the transfer of solute molecules across a semi-permeable membrane.
B Solvent molecules can move from a dilute solution to a concentrated solution.
C Solute molecules can move from a concentrated solution to a dilute solution.
D Osmosis between solutions continues until the mole fractions of solute and solvent are equal.

15 Which one of the following solutions has an osmotic pressure greater than 1 atm at 0 °C?

A 0.1 mole glucose in 2.24 dm^3 of aqueous solution
B 0.5 mole of sodium chloride (NaCl) in 22.4 dm^3 of aqueous solution
C 1 mole potassium iodide in 22.4 dm^3 of aqueous solution
D 2 moles sucrose in 44.8 dm^3 of aqueous solution

16 Which one of the following 0.01 M aqueous solutions has the greatest osmotic pressure?

A NH_3
B KOH
C H_2SO_4
D CH_3COOH

17. 24 g of compound X, dissolved in 11.2 dm^3 of hexane causes an osmotic pressure of 0.5 atm at 0 °C. What is the relative molecular mass of X?

A 12
B 24
C 48
D 96

18 Which one of the following pairs of liquid mixtures would you expect to show behaviour nearest to ideal?

A methanol + water
B propanone + ethanol
C 2-methylpentane + hexane
D trichloromethane + propanone

19 A monobasic acid HA is dissociated in a non-aqueous solution:

$$HA \rightleftharpoons H^+ + A^-$$

1 mole of HA produces a depression of freezing point of 0.4 °C if dissolved in 1 kg of solvent. The cryoscopic constant for the solution is 0.3 °C kg^{-1}. What is the degree of dissociation of HA?

A $\dfrac{0.4 - 0.3}{0.4}$

B $\dfrac{0.4}{0.3}$

C $\dfrac{0.3}{0.4}$

D $\dfrac{0.4}{0.3} - 1$

IX Acid/Base Reactions

1 Which one of the following expressions is *not* numerically equal to the pH of a solution at 25 °C?

A $-\log_{10}[H^+]$

B $\log_{10}\dfrac{1}{[H^+]}$

C $\dfrac{1}{\log_{10}[H^+]}$

D $14+\log_{10}[OH^-]$

2 How could you cause the pH of a solution to rise from three to six?

A   double the hydrogen ion concentration
B   half the hydrogen ion concentration
C   decrease the hydrogen ion concentration one thousand times
D   decrease the hydroxide ion concentration one thousand times

3 The hydrogen ion concentration of a solution is increased one hundred times. Which one of the following pH changes could occur?

	Initial pH	Final pH
A	1	2
B	1	3
C	4	2
D	6	3

4 The ionic product of water at 70 °C is $1\times10^{-12}\,\text{mol}^2\,\text{dm}^{-6}$. A correct statement about water at this temperature is that

A $pH=7$.
B $[H^+]=1\times10^{-6}\,\text{mol dm}^{-3}$.
C $[OH^-]>[H^+]$.
D $[H^+]>[OH^-]$.

5 The ionic products of water at 25 °C and 30 °C are $1\times10^{-14}\,\text{mol}^2\,\text{dm}^{-6}$ and $4\times10^{-14}\,\text{mol}^2\,\text{dm}^{-6}$ respectively. From this information determine which one of the following statements is *incorrect*.

A   The dissociation of water is an endothermic process.
B $[OH^-]$ at 30 °C is greater than $[OH^-]$ at 25 °C in pure water.
C   The pH of water at 25 °C is 7.
D   The pH of water at 30 °C is greater than 7.

6 The pH of a solution of a fully ionized, monobasic acid is −1. What would you expect the concentration of hydrogen ions in the solution to be?

A   0.01 mol dm^{-3}
B   0.1 mol dm^{-3}
C   1.0 mol dm^{-3}
D   10.0 mol dm^{-3}

7 A weak monobasic acid is 10% dissociated in 0.01 M aqueous solution. Its likely pH value is

A   0.1.
B   1.0.
C   2.0.
D   3.0.

8 Concentrated nitric acid and concentrated sulphuric acid react together according to the equation:

$$2\,H_2SO_4 + HNO_3 \rightarrow 2\,HSO_4^- + NO_2^+ + H_3O^+$$

Which one of the following is a true statement about this reaction?

A   Sulphuric acid is acting as an acid.
B   Nitric acid is acting as an acid.
C   Sulphuric acid is acting as a dehydrating agent.
D   Nitric acid is acting as an oxidizing agent.

9 49 cm^3 of 0.1 M potassium hydroxide solution is added to 50 cm^3 of 0.1 M nitric acid. The approximate pH of the resulting solution will be

A   1.
B   3.
C   6.
D   7.

10 Which one of the following gives an acidic solution when dissolved in water?

A   ammonia
B   sodium chloride
C   sodium carbonate
D   aluminium chloride

11 The K_a values for propanoic acid and trichloroethanoic acid are 1×10^{-5} mol dm^{-3} and 1×10^{-1} mol dm^{-3} respectively. What is the value of the quotient

$$\frac{pK_a \text{ (propanoic acid)}}{pK_a \text{ (trichloroethanoic acid)}}?$$

A 10^{-4}
B   0.2
C   5
D 10^4

12 What is the hydrogen ion concentration in mol dm^{-3} of an aqueous solution of sulphuric acid which has a pH of 2?

A 5×10^{-3}
B 1×10^{-2}
C 2×10^{-2}
D 5×10^{-2}

13 Trichloroethanoic acid ionizes in aqueous solution according to the equation:
$$Cl_3CCOOH(s) + H_2O(l) \rightleftharpoons Cl_3CCOO^-(aq) + H_3O^+(aq)$$

The pK_a value for the acid is

A $\dfrac{[Cl_3CCOO^-][H_3O^+]}{[Cl_3CCOOH]}$

B $-\log_{10} \dfrac{[Cl_3CCOO^-][H_3O^+]}{[Cl_3CCOOH][H_2O]}$

C $\log_{10} \dfrac{[Cl_3CCOOH]}{[Cl_3CCOO^-][H_3O^+]}$

D $-\log_{10} \dfrac{[Cl_3CCOOH]}{[Cl_3CCOO^-][H_3O^+]}$

14 In the reactions shown below, in what way(s) does $NH_3(aq)$ act?

(a) $NH_3(aq) + H_2O(l) \rightleftharpoons NH_4^+(aq) + OH^-(aq)$
(b) $2\,NH_3(aq) \rightleftharpoons NH_4^+(aq) + NH_2^-(aq)$

A acid and base
B oxidizing agent and base
C base and reducing agent
D acid only

15 What is the pH of 0.001 M sodium hydroxide solution at 25 °C?

A 15
B 14
C 11
D 3

16 Which one of the following assumptions must be made in order to answer question 15?

A The ionization of water is an endothermic process.
B A 0.001 M aqueous solution of sodium hydroxide is completely ionized.
C The ionic product of water is temperature dependent.
D The solution contains no hydrogen ions.

17 Which one of the following would be the least effective buffering system?

A CH_3COOH/CH_3COONa
B $HCl/NaCl$
C NH_3/NH_4Cl
D $HCOOH/HCOONa$

18 A correct statement about a solution of hydrogen chloride in water is that

A in all solutions, the hydrogen chloride is fully dissociated into ions.
B the minimum pH of such solutions is zero.
C the pH of the solution is dependent upon the degree of ionization of the solute as well as the concentration of the solution.
D the degree of ionization of the solute is independent of the concentration of the solution.

19 A weak dibasic acid can dissociate in two stages:

(i) $H_2A \rightleftharpoons HA^- + H^+$
(ii) $HA^- \rightleftharpoons A^{2-} + H^+$

A dissociation constant can be written for each stage:

$$K_i = \frac{[HA^-][H^+]}{[H_2A]} \qquad K_{ii} = \frac{[A^{2-}][H^+]}{[HA^-]}$$

The overall dissociation reaction is:

$$H_2A \rightleftharpoons 2H^+ + A^{2-}.$$

If the dissociation constant for the overall reaction is designated K_a, what is the relationship between K_a, K_i and K_{ii}?

A $K_a = K_i + K_{ii}$

B $K_a = \dfrac{K_{ii}}{K_i}$

C $K_a = K_i \times K_{ii}$

D $K_a = \dfrac{K_i}{K_{ii}}$

20 The value of K_a for ethanoic acid is 1.8×10^{-5} mol dm^{-3} whilst that for phenol is 1.2×10^{-10} mol dm^{-3}. Which one of the following statements can be correctly deduced from this information?

A Equimolar solutions of the two substances will have equal pH values.
B Phenol is the stronger of the acids.
C Phenol has a larger pK_a value than ethanoic acid.
D A 0.1 M solution of phenol in water will have a pH of 7.

X Redox Reactions

1 In which one of the following does chromium have an oxidation number other than six?

    A CrO_4^{2-}
    B $Cr_2O_7^{2-}$
    C CrO_3
    D $Cr(OH)_3$

2 In which one of the following changes are five electrons transferred?

    A $MnO_4^- \rightarrow Mn^{2+}$
    B $Cr_2O_7^{2-} \rightarrow 2\,Cr^{3+}$
    C $MnO_4^{2-} \rightarrow MnO_2$
    D $CrO_4^{2-} \rightarrow Cr^{3+}$

3 Which one of the following reactions involves neither oxidation nor reduction?

    A $MnO_4^-(aq) + 5\,Fe^{2+}(aq) + 8\,H^+(aq) \rightarrow Mn^{2+}(aq) + 5\,Fe^{3+}(aq) + 4\,H_2O(l)$
    B $2\,KI(aq) + Cl_2(aq) \rightarrow 2\,KCl(aq) + I_2(aq)$
    C $2\,CrO_4^{2-}(aq) + 2\,H^+(aq) \rightarrow Cr_2O_7^{2-}(aq) + H_2O(l)$
    D $2\,H_2SO_4(l) + Cu(s) \rightarrow CuSO_4(aq) + 2\,H_2O(l) + SO_2(g)$

4 Which one of the following reactions does *not* involve disproportion?

    A $2\,TiCl_3(aq) \rightarrow TiCl_2(aq) + TiCl_4(aq)$
    B $3\,MnO_4^{2-}(aq) + 4\,H^+(aq) \rightarrow 2\,MnO_4^-(aq) + MnO_2(s) + 2\,H_2O(l)$
    C $2\,KOH(aq) + Cl_2(g) \rightarrow KCl(aq) + KOCl(aq) + H_2O(l)$
    D $2\,H_2SO_4(l) + Cu(s) \rightarrow CuSO_4(aq) + 2\,H_2O(l) + SO_2(g)$

5 The reaction between manganate(VII) ions and ethanedioate ions in acidic solution can be summarized by:

$$C_2O_4^{2-}(aq) \rightarrow 2\,CO_2(g) + 2\,e^-$$
$$8\,H^+(aq) + MnO_4^-(aq) + 5\,e^- \rightarrow Mn^{2+}(aq) + 4\,H_2O(l)$$

How many moles of the ethanedioate ions are oxidized by one mole of the manganate(VII) ions?

    A 0.4
    B 1.0
    C 2.5
    D 5.0

6 Concentrated sulphuric acid can be reduced by

    A NaCl.
    B NaOH.
    C $NaNO_3$.
    D NaBr.

7 If an acidic solution of hydrogen peroxide is treated with potassium manganate(VII) solution

    A   the manganate(VII) is reduced to manganese(IV) oxide.
    B   the hydrogen peroxide is converted to hydrogen.
    C   the manganate(VII) is converted into manganese(III).
    D   the hydrogen peroxide is oxidized to oxygen.

8 In which of the following reactions is hydrogen acting as an oxidizing agent?

    A $Ca(s) + H_2(g) \rightarrow CaH_2(s)$
    B $F_2(g) + H_2(g) \rightarrow 2\,HF(g)$
    C $C_2H_2(g) + H_2(g) \rightarrow C_2H_4(g)$
    D $O_2(g) + 2\,H_2(g) \rightarrow 2\,H_2O(l)$

9 The oxidation state of iron remains unchanged when

    A   magnesium ribbon is added to iron(II) sulphate solution.
    B   hydrogen sulphide gas is passed through iron(III) chloride solution.
    C   hydrogen is passed over heated $Fe_3O_4(s)$.
    D   sodium hydroxide solution is added to iron(III) nitrate solution.

10 Which one of the following compounds can be further oxidized?

    A $SO_3(g)$
    B $PbO_2(s)$
    C $ClO_2(g)$
    D $MgO(s)$

11 Which one of the following changes in oxidation state occurs during the reaction $2\,Cu^{2+}(aq) + 4\,I^-(aq) \rightarrow I_2(aq) + 2\,CuI(s)$?

    A   Cu(O) → Cu(I)
    B   I(−I) → I(0)
    C   I(0) → I(I)
    D   Cu(II) → Cu(IV)

12 Given the following $E^\ominus$ values:

$$K^+(aq)|K(s); \quad -2.92\ V$$
$$Zn^{2+}(aq)|Zn(s); \quad -0.76\ V$$
$$Fe^{2+}(aq)|Fe(s); \quad -0.44\ V$$
$$Ag^+(aq)|Ag(s); \quad +0.80\ V$$

Which one of the following is correct?

    A   Silver metal is the most powerful reducing agent listed.
    B   Zinc metal should be able to be used to give a sacrificial protective coating to iron.
    C   Potassium metal is the strongest oxidizing agent listed.
    D   Iron cannot displace silver from a solution of silver ions in water.

13 When the cell shown in the diagram below is set up which one of the following changes does *not* take place?

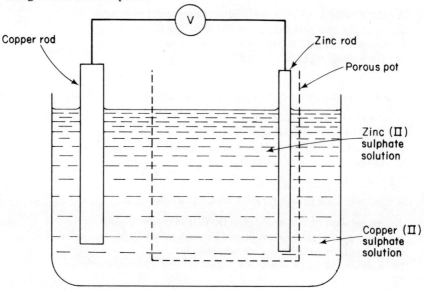

A   Copper is deposited on to the surface of the copper electrode.
B   Electrons are transferred from the zinc to the copper electrode.
C   Zinc metal is deposited on to the surface of the zinc electrode.
D   Sulphate ions migrate towards the zinc electrode.

14 The following cell is set up:
$$Zn(s)|Zn^{2+}(aq)|Cu^{2+}(aq)|Cu(s)$$
The $E^{\ominus}$ values for the half-cells are:
$$Zn^{2+}(aq)|Zn(s); \quad -0.76 \text{ V}$$
$$Cu^{2+}(aq)|Cu(s); \quad +0.34 \text{ V}$$
What is the e.m.f. for the cell as it is written above?

A -1.10 V
B -0.42 V
C $+0.42$ V
D $+1.10$ V

15 Which one of the following *cannot* be correctly deduced about the reaction between manganate(VII) ions and hydrogen sulphide solution from the following half-cell diagrams?
$$[2 \text{ H}^+(aq) + S(s)], H_2S(aq)|Pt; \quad E^{\ominus} = +0.14 \text{ V}$$
$$[MnO_4^-(aq) + 8 \text{ H}^+(aq)], [Mn^{2+}(aq) + 4 \text{ H}_2O(l)]|Pt; E^{\ominus} = +1.51 \text{ V}$$

A   The exact equilibrium position of the reaction is pH dependent.
B   In acidic solution, the equilibrium lies on the products side.
C   In the reaction, the manganese is oxidized and the sulphur is reduced.
D   In acidic solution, manganate(VII) is a more effective oxidizing agent than sulphur.

In question 16 to 18 you are shown several cell diagrams involving the metals L, M, N, Q and R and their respective two-positive ions, $L^{2+}(aq)$, etc. The standard electrode potentials of each metal/metal ion system is shown here:

Half-cell	e.m.f./volts
$L^{2+}(aq)\|L(s)$	-1.2
$M^{2+}(aq)\|M(s)$	-0.8
$N^{2+}(aq)\|N(s)$	-0.6
$Q^{2+}(aq)\|Q(s)$	-0.2
$R^{2+}(aq)\|R(s)$	$+0.4$

Using the responses A to D below, determine the potentials of the standard cells as written in questions 16 to 18. Each response may be used once, more than once or not at all.

A +0.2 V
B +0.6 V
C +1.0 V
D +1.6 V

16 $L(s)\|L^{2+}(aq)\vdots R^{2+}(aq)\|R(s)$

17 $M(s)\|M^{2+}(aq)\vdots N^{2+}(aq)\|N(s)$

18 $Q(s)\|Q^{2+}(aq)\vdots R^{2+}(aq)\|R(s)$

19 Which one of the following statements about the Daniell cell:

$$Zn(s)\|Zn^{2+}(aq)\vdots Cu^{2+}(aq)\|Cu(s)$$

is correct?

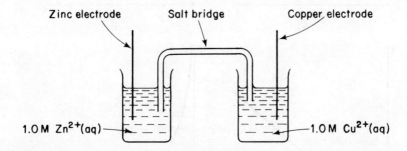

Lowering the concentration of zinc ions in the zinc half cell

A will affect the potential of the copper half-cell.
B will make the potential of the zinc half-cell more positive.
C will make the potential of the cell, as written in the question, more negative.
D and raising the concentration of the copper ions in the copper half-cell will necessarily alter the potential of the whole cell.

20 A standard electrode system which has a greater tendency to release electrons than the standard hydrogen electrode is

  A $Cu^{2+}(aq)|Cu(s)$.
  B $Ag^+(aq)|Ag(s)$.
  C $Zn^{2+}(aq)|Zn(s)$.
  D $Hg^{2+}(aq)|Hg(l)$.

21 The Nernst equation is

$$E^\ominus = \frac{2.3RT\log_{10}K_c}{zF}$$

where

  R is the gas constant/J K^{-1} mol^{-1}
  T is the temperature/K
  z is the number of electrons transferred in the reaction
  F is the Faraday constant/coulombs mol^{-1}

and

  K_c is the equilibrium constant

Which one of the following graphs represents a plot of $E^\ominus$/volts (y axis) against $\log_{10} K_c$ (x axis) at constant temperature?

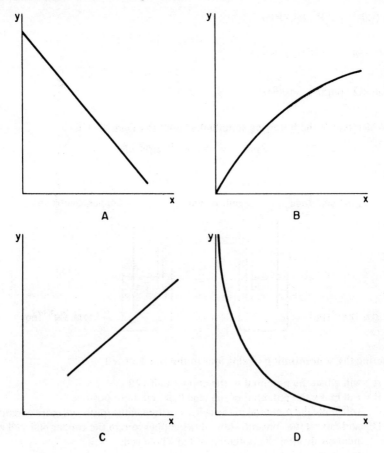

A

B

C

D

XI Groups I, II and III

1 Which one of these elements is the least reactive towards water?

    A Be
    B Ca
    C Sr
    D Ba

2 Which one of these hydrides will react with water to produce hydrogen?

    A HF
    B H_2S
    C CH_4
    D NaH

3 Which one of the following is least soluble in water?

    A LiF
    B NaF
    C RbF
    D CsF

4 Which one of the following molecular formulae is *not* known for a Group I oxide?

    A M_2O
    B MO
    C MO_2
    D M_2O_2

5 Which one of the following is the strongest base?

    A LiOH
    B NaOH
    C KOH
    D RbOH

6 Which one of the following is the most thermally stable at 1000 °C?

    A lithium carbonate
    B sodium hydrogencarbonate
    C potassium carbonate
    D calcium hydrogencarbonate

7 Which one of the following produces a solid nitrogen-containing compound when heated strongly?

  A $LiNO_3$
  B $NaNO_3$
  C $Ca(NO_3)_2$
  D $Mg(NO_3)_2$

8 When $[Al(H_2O)_6]^{3+}$ undergoes hydrolysis it forms a series of complex ions of formula $[Al(H_2O)_x(OH)_y]^{z+}$. Which one of the following pairs of equations represents the relationships between x, y and z?

  A $x + y = 6;$ $z = 3 - x$
  B $x - y = 4;$ $z = 6 - x$
  C $x + y = 6;$ $z = 3 - y$
  D $y - x = 6;$ $z = 3 - y$

9 Which one of the following statements about aluminium halides is correct?

  A Aluminium bromide is insoluble in organic solvents.
  B Aluminium fluoride is hydrolyzed by water.
  C Aluminium chloride can exist as a dimer, Al_2Cl_6, below 400 °C.
  D Aluminium fluoride forms a complex ion $[AlF_6]^{6-}$.

10 Which one of the following is an *incorrect* statement about the reversible reaction represented by the following equation?

$$[Al(H_2O)_6]^{3+}(aq) \rightleftharpoons [Al(H_2O)_5(OH)]^{2+}(aq) + H^+(aq)$$

  A $[Al(H_2O)_6]^{3+}(aq)$ is acting as an acid.
  B The ionization is brought about by the high polarising power of Al^{3+}.
  C Addition of sodium carbonate moves the position of equilibrium to the left.
  D $[Al(H_2O)_6]^{3+}$ does not undergo hydrolysis in a strongly acidic solution.

11 Which one of the following statements about the chemistry of aluminium chloride is correct?

  A The bonding is predominantly ionic.
  B Above 800 °C the molecule is trigonal pyramidal in shape.
  C At room temperature the aluminium atom in solid aluminium chloride has a co-ordination number of three.
  D It undergoes hydrolysis to form the $[Al(H_2O)_6]^{3+}(aq)$ ion.

12 Which one of the following is produced by hydrolysis of BCl_3?

  A $[B(H_2O)_6]^{3+}$
  B H_3BO_3
  C B_2O_3
  D $[B(OH)_6]^{3-}$

13 Why can you *not* detect the presence of magnesium by a flame test using a Bunsen burner?

    A   There are no excited states in the magnesium atom.

    B   The Bunsen flame is not hot enough to excite the electrons in the magnesium atom.

    C   Once a magnesium electron has been excited it will remain in the excited state.

    D   The radiation emitted by electronic transitions within the magnesium atom cannot be seen by the human eye.

14 The maximum number of electron pairs which can be found in the lithium atom is

    A   0.

    B   1.

    C   2.

    D   3.

15 Group I elements have similar chemical properties because

    A   all the elements crystallize in body-centred cubic structures.

    B   they all have the electronic configuration ns^1.

    C   they all give rise to mainly ionic compounds.

    D   their electron affinities are very low.

16 In which one of the following ways are lithium and magnesium most alike?

    A   Their electronic configurations are similar.

    B   Their electronegativities are both high.

    C   They have approximately the same atomic radius.

    D   Their compounds with chlorine have the same molar ratio of metal : non-metal.

17 The reaction represented by the equation

$$CaH_2(s) + 2\,H_2O(l) \rightarrow Ca(OH)_2(aq) + 2\,H_2(g)$$

involves

    A   the oxidation of calcium.

    B   the reduction of calcium.

    C   an acid/base reaction.

    D   the oxidation and reduction of hydrogen.

18 Which one of the following pairs of ions could you use to make an alum of formula $XY(SO_4)_2 \cdot 12H_2O$?

	X	Y
A	Na^+	Cr^{3+}
B	Na^+	Mg^{2+}
C	Li^+	Al^{3+}
D	Mg^{2+}	Fe^{3+}

Questions 19 to 21 refer to four solutions of Group I and II compounds listed below as responses A to D.

A   2 dm³ of 2 M KBr(aq)
B   1 dm³ of 1 M CaCl₂(aq)
C   2 dm³ of 0.5 M MgCl₂(aq)
D   1 dm³ of 2.5 M Na₂SO₄(aq)

In answering questions 19 to 21 you may use each of the responses once, more than once or not at all.

Which solution contains

19 the greatest number of ions?

20 the greatest concentration of cations?

21 the smallest concentration of anions?

XII Groups IV, V and VI

1 Which one of the following Group IV chlorides is *not* hydrolyzed?

    A $SnCl_4$
    B $GeCl_4$
    C $SiCl_4$
    D CCl_4

2 Which one of these elements is the most stable in the $+2$ oxidation state?

    A   Si
    B   Ge
    C   Sn
    D   Pb

3 Which one of these statements concerning Group IV elements is correct?

    A   The stability of compounds with oxidation state $+4$ increases with increasing atomic size.
    B   Only tin and lead are capable of forming M^{2+} ions in stable compounds.
    C   Only tin and lead can form complexes of the type $[MF_6]^{2-}$.
    D   Lead(IV) compounds have strong reducing properties.

4 Which one of the following elements does *not* produce an insoluble oxide when treated with cold concentrated nitric acid?

    A   Si
    B   Ge
    C   Sn
    D   Pb

5 Which one of the following is a correct description of the oxidation state of tin and the shape of the complex ion formed by treating tin(IV) chloride with excess concentrated hydrochloric acid?

	Oxidation state	Shape
A	4	octahedral
B	2	tetrahedral
C	6	octahedral
D	4	square planar

6 Which one of the following halides has the greatest ionic character?

    A SiF_4
    B $PbCl_4$
    C SnI_4
    D PbF_4

7 Which one of the following reactions may be used to prepare $SiCl_4$?

    A Si + concentrated $HCl(aq)$ + heat
    B Si + $Cl_2(g)$ + heat
    C SiO_2 + $Cl_2(g)$ + heat
    D SiO_2 + concentrated $HCl(aq)$ + heat

8 Which one of the following contains an odd number of electrons?

    A NO_3^-
    B NO_2
    C N_2O_4
    D NO_2^+

9 Which one of the following statements about Group V elements is correct?

    A There is a decrease in metallic character down the group.
    B The thermal stability of the hydrides, MH_3, decreases down the group.
    C The oxides become increasingly acidic down the group.
    D All the elements can form tripositive cations in solution.

10 Which one of the following statements comparing ammonia and phosphine is correct?

    A Ammonia has lower melting and boiling points than phosphine.
    B Ammonia is less soluble in water than phosphine.
    C Phosphine is a more powerful reducing agent than ammonia.
    D Ammonia is less thermally stable than phosphine.

11 Ammonia *cannot* be produced by

    A heating ammonium chloride with sodium hydroxide.
    B treating magnesium nitride with water.
    C thermal decomposition of ammonium nitrite (NH_4NO_2).
    D warming ethanamide with potassium hydroxide.

12 Which one of these ions when mixed with ammonia in aqueous solution produces a precipitate which will *not* dissolve when excess ammonia solution is added?

    A Fe^{3+}
    B Zn^{2+}
    C Cu^{2+}
    D Ag^+

13 Which one of the following is a correct statement concerning the shape of an ammonia molecule?

    A It is a V-shaped molecule.
    B Its shape is based on a tetrahedron with lone pairs in two positions.
    C The H–N–H angles are less than 109° 28′.
    D The repulsion between the lone pair and bonding pairs is less than that between the bonding pairs.

14 Nitrogen has an oxidation number of -1 in

A NO_3^-.
B N_2O.
C NH_2OH.
D HNO_2.

15 In which one of the following reactions is hydrogen peroxide acting as a reducing agent?

A $PbS(s) + 4 H_2O_2(aq) \rightarrow PbSO_4(aq) + 4 H_2O(l)$
B $2[Fe(CN)_6]^{3-}(aq) + H_2O_2(aq) + 2 OH^-(aq) \rightarrow$
$$2[Fe(CN)_6]^{4-}(aq) + 2 H_2O(l) + O_2(g)$$
C $2 I^-(aq) + 2 H^+(aq) + H_2O_2(aq) \rightarrow I_2(aq) + 2 H_2O(l)$
D $2 Fe^{2+}(aq) + H_2O_2(aq) + 2 H^+(aq) \rightarrow 2 Fe^{3+}(aq) + 2 H_2O(l)$

16 In which one of the following reactions is hydrogen sulphide acting as a reducing agent?

A $2 OH^-(aq) + H_2S(aq) \rightarrow 2 H_2O(l) + S^{2-}(aq)$
B $Cu^{2+}(aq) + H_2S(aq) \rightarrow CuS(s) + 2 H^+(aq)$
C $H_2SO_4(aq) + H_2S(aq) \rightarrow SO_2(g) + 2 H_2O(l) + S(s)$
D $H_2O(l) + H_2S(aq) \rightarrow H_3O^+(aq) + HS^-(aq)$

17 Which one of the following has the lowest boiling point at 1 atm pressure?

A H_2O
B H_2S
C H_2Se
D H_2Te

18 Which one of these combinations of bonds would nitrogen $(1s^2 2s^2 2p^3)$ normally form in its ground state?

A 3 covalent and 1 dative
B 2 covalent and 1 dative
C 1 covalent and 1 dative
D 1 covalent and 2 dative

19 Phosphorus $(1s^2 2s^2 2p^6 3s^2 3p^3)$ can give rise to five bonds if sufficient energy is available to promote an electron to a higher level. Which one of these combinations of bonds would be formed?

A 2 covalent and 3 dative
B 3 covalent and 2 dative
C 4 covalent and 1 dative
D 5 covalent

20 Which one of the following conversions is neither an oxidation nor a reduction?

A $SO_2 \rightarrow SO_4^{2-}$
B $2 S \rightarrow S_2O_3^{2-}$
C $SO_2 \rightarrow HSO_3^-$
D $S_2O_3^{2-} \rightarrow S_2O_4^{2-}$

XIII Group VII

1 Astatine

    A   has a larger first ionization energy than chlorine.
    B   has a smaller atomic radius than bromine.
    C   is more electronegative than iodine.
    D   gives rise to an ion (At^-) which has a larger radius than the fluoride ion.

2 Chlorine free radicals ($Cl\cdot$)

    A   combine together to form a chlorine molecule and electrons.
    B   remove hydrogen from hydrocarbons giving rise to hydrogen chloride.
    C   can contain either an odd number of electrons or an even number.
    D   are formed by the action of infra-red radiation on chlorine molecules.

3 Which one of the following is the strongest oxidizing agent?

    A   iodine
    B   bromine
    C   chlorine
    D   fluorine

4 Disproportionation is *not* involved in the reaction:

    A $2\,F_2(g) + 2\,H_2O(l) \rightarrow 4\,HF(g) + O_2(g)$
    B $Cl_2(g) + H_2O(l) \rightarrow HCl(aq) + HOCl(aq)$
    C $Cl_2(g) + 2\,KOH(aq) \rightarrow KCl(aq) + KOCl(aq) + H_2O(l)$
    D $Br_2(l) + H_2O(l) \rightarrow HBr(aq) + HOBr(aq)$

5 Which one of the following compounds has the highest boiling point?

    A   hydrogen fluoride
    B   hydrogen chloride
    C   hydrogen bromide
    D   hydrogen iodide

6 Fluorine will react with

    A SF_6.
    B PF_5.
    C CF_4.
    D   Xe.

7 Which one of the following halogens has the lowest heat of atomization?

    A   fluorine
    B   chlorine
    C   bromine
    D   iodine

8 Which one of the following quantities does *not* contribute to the conversion $\frac{1}{2}$ $Cl_2(g) \rightarrow Cl^-(aq)$?

    A   the enthalpy change of atomization of chlorine
    B   the ionization energy of a gaseous chlorine atom
    C   the electron affinity of a gaseous chlorine atom
    D   the hydration energy of a gaseous chloride ion

9 A soluble silver(I) salt can be formed by mixing silver(I) nitrate with an aqueous solution of

    A   fluoride ions.
    B   chloride ions.
    C   bromide ions.
    D   iodide ions.

10 Which one of the following reactions will *not* give rise to a halogen?

    A   sodium iodide + concentrated sulphuric acid
    B   sodium chloride + manganese(IV) oxide + concentrated sulphuric acid
    C   sodium fluoride + concentrated sulphuric acid
    D   sodium bromide + concentrated sulphuric acid

11 Chlorine gas will react with

    A   sodium fluoride solution.
    B   sodium iodide solution.
    C   calcium chloride crystals.
    D   silicon tetrachloride.

12 Solid silver chloride is most soluble in

    A   0.1 M HCl (aq).
    B   0.1 M NH_3(aq).
    C   0.1 M NaCl(aq).
    D   water.

13 Iodine molecules, I_2(aq), react with thiosulphate(VI) ions, $S_2O_3^{2-}$(aq), to form iodide ions, I^-(aq), and tetrathionate ions, $S_4O_6^{2-}$(aq). The number of moles of thiosulphate(VI) ions that will react with one mole of iodine is

    A   1.
    B   2.
    C   3.
    D   4.

14 Which one of the following hydrogen halides is energetically the least stable with respect to decomposition into its elements?

    A   HF(g)
    B   HCl(g)
    C   HBr(g)
    D   HI(g)

15 Which of the following gases are produced by adding concentrated sulphuric acid to potassium chloride in the cold?

    A Cl_2, SO_2, HCl.
    B Cl_2, H_2S.
    C Cl_2, SO_2.
    D HCl only.

16 The most electronegative halogen is

    A fluorine.
    B chlorine.
    C bromine.
    D iodine.

17 When preparing a sample of dry chlorine gas by the reaction of concentrated hydrochloric acid with manganese(IV) oxide, the gas formed is bubbled through two wash bottles. These bottles should contain:

	First bottle	Second bottle
A	concentrated sulphuric acid	water
B	concentrated sulphuric acid	concentrated sulphuric acid
C	water	concentrated sulphuric acid
D	water	water

XIV Transition Metal Chemistry

1 Which one of the following elements can exhibit the highest oxidation number?

    A Ti
    B V
    C Mn
    D Co

2 Which one of the following compounds includes a transition metal in oxidation state zero?

    A $[Co(NH_3)_6]Cl_3$
    B $[Fe(H_2O)_6]SO_4$
    C $[Ni(CO)_4]$
    D $[Fe(H_2O)_3(OH)_3]$

3 An example of a complex ion which is oxidized by air is

    A $[Fe(H_2O)_6]^{3+}$
    B $[Co(NH_3)_6]^{2+}$
    C $[Cu(H_2O)_6]^{2+}$
    D $[Ag(NH_3)_2]^{+}$

4 When copper(I) sulphate is added to water the products will include

    A $CuSO_4(aq)$ and $Cu(OH)_2(s)$.
    B $Cu(s)$ and $[Cu(H_2O)_6]^{2+}(aq)$.
    C $Cu(s)$ and $CuOH(s)$.
    D $Cu(OH)_2(s)$ and $Cu(s)$.

5 Which one of the following statements about the chemistry of chromium is true?

    A The reaction $2\ CrO_4^{2-}(aq) + 2\ H^+(aq) \rightarrow Cr_2O_7^{2-}(aq) + H_2O(l)$ involves the oxidation of chromium.
    B Common oxidation states of chromium include $+2$, $+3$, $+4$ and $+5$.
    C Dichromate(VI) oxidizes iron(II) and is itself reduced to chromium(II).
    D Chromate(VI) can be prepared by oxidizing a chromium(III) salt with sodium peroxide (Na_2O_2).

6 What is the co-ordination number of zinc in the complex ion $[Zn(OH)_4(H_2O)_2]^{2-}$?

    A 0
    B 2
    C 4
    D 6

7 Which one of the following correctly describes the change in colour and geometry when concentrated hydrochloric acid is added to a solution of $[Co(H_2O)_6]^{2+}(aq)$?

	Colour of solution	Geometry of the complex ion
A	pink—no change	octahedral → square planar
B	pink → blue	octahedral → tetrahedral
C	blue → green	octahedral—no change
D	pink → blue	tetrahedral → octahedral

8 When iron(II) sulphate is dissolved in water and an aqueous solution of potassium cyanide KCN(aq) is added, the complex formed is

A   tetrahedral and anionic.
B   linear and cationic.
C   octahedral and anionic.
D   square planar and cationic.

9 On adding potassium iodide solution to copper(II) sulphate solution, the copper salt formed is

A CuI_2.
B CuI.
C Cu_2SO_4.
D $CuSO_3$.

10 The electronic arrangement in Cr^{3+} ions is

A $(Argon)3d^3$.
B $(Argon)3d^1\ 4s^2$.
C $(Argon)3d^2\ 4s^1$.
D $(Argon)3d^1\ 4p^2$.

11 A black oxide of manganese (X) is fused with potassium hydroxide and potassium nitrate. Extraction with water yields a green solution (Y). Addition of chlorine to the green solution produces a purple solution (Z). The oxidation states of manganese in X, Y and Z are:

	Oxidation states		
	X	Y	Z
A	2	6	7
B	4	6	6
C	4	6	7
D	4	7	6

12 Dissolving copper(II) oxide in excess concentrated hydrochloric acid produces

A $CuCl(s)$.
B $CuCl_2 \cdot 2H_2O(s)$.
C $[CuCl_4]^{2-}(aq)$.
D $[Cu(H_2O)_6]^{2+}(aq)$.

13 Which one of the following is *not* isoelectronic with the remaining species?

   A Fe^{3+}
   B Mn^{2+}
   C Sc^{3+}
   D   V

14 The complex ion $[MX_2Y_2]^{2-}$ is square planar. X and Y are singly charged ligands and the metal (M) is in oxidation state $+2$. How many geometrical isomers does the complex have?

   A   0
   B   2
   C   3
   D   4

15 The complex $[MX_2Y_2]$ is octahedral, where X is a monodentate ligand and Y a bidentate ligand. For this complex there will be

   A   no isomers.
   B   two geometrical isomers, neither of which is optically active.
   C   two geometrical isomers, both of which are optically active.
   D   three geometrical isomers, two of which are optically active.

16 The ligand $NH_2-CH_2-CH_2-NH_2$ can complex to a metal atom through

   A   one atom.
   B   two atoms.
   C   three atoms.
   D   four atoms.

17 How many moles of silver chloride will be precipitated if excess silver nitrate solution is added to a solution containing one mole of $[CrCl(H_2O)_5]Cl_2$?

   A   0
   B   1
   C   2
   D   3

18 In alkaline solution the conversion $MnO_4^-(aq) \rightarrow MnO_3^-(aq)$ is possible. This is a reduction. How many moles of electrons are required to reduce one mole of $MnO_4^-(aq)$ in this way?

   A   1
   B   2
   C   3
   D   5

XV Fundamental Organic Chemistry

1 A hydrocarbon has a density of 2.5 g dm^{-3} at $0\,^\circ\text{C}$, 1 atm. Which one of the following is the molecular formula of the hydrocarbon? The volume of one mole of a gas at s.t.p. is 22.4 dm^3. (A_r values: $C = 12$, $H = 1$.)

A C_2H_2
B C_2H_4
C C_4H_8
D C_4H_{10}

2 Complete combustion of 0.448 dm^3 of a volatile hydrocarbon (measured at s.t.p.) produced 0.06 mole of carbon dioxide. How many carbon atoms are there in one molecule of the hydrocarbon?

A 2
B 3
C 6
D 12

3 A compound has an empirical formula CH_2. Its vapour density is 56. What is the molecular formula of the compound?

A CH_2
B C_2H_4
C C_4H_8
D C_8H_{16}

4 0.42 g of a hydrocarbon occupies 0.112 dm^3 at s.t.p. Which one of the following is its empirical formua? (A_r values: $C = 12$, $H = 1$.)

A CH_2
B CH_3
C C_2H_4
D C_2H_5

5 Which one of the following is an example of homolytic fission?

A $C_5H_{12} \rightarrow C_2H_5^{\cdot} + C_3H_7^{\cdot}$
B $CH_3CH_2^+ \rightarrow C_2H_4 + H^+$
C $C_2H_5Br + OH^- \rightarrow C_2H_5OH + Br^-$
D $CH_3CH_2Br \rightarrow C_2H_4 + HBr$

6 Which one of the following *cannot* act as a nucleophile?

A H_2O
B NH_3
C CN^-
D NO_2^+

7 In which one of the following compounds are sp^2 hybrid orbitals *not* involved in the bonding?

A but-1-ene
B propanone
C ethanol
D methanal

8 Which one of the following exhibits optical isomerism?

A $HOCH_2NH_2$
B $(CH_3)_2C(OH)COOH$
C $CH_3CH(OH)COOH$
D $C_2H_5CH{=}CHCH_3$

9 Complete combustion of $1 \, dm^3$ of a hydrocarbon required $8 \, dm^3$ of oxygen, measured at the same temperature and pressure. The molecular formula of the hydrocarbon could be

A C_3H_8.
B C_4H_8.
C C_5H_{12}.
D C_8H_{16}.

10 Which one of the following bonds is the least polar?

A C—H in propane
B C—Cl in chlorobenzene
C C—C in 1,1-dichloroethane
D C—C in ethane

11 Which one of the following acids will have the highest pK_a value?

A propanoic acid
B ethanoic acid
C chloroethanoic acid
D trichloroethanoic acid

12 Which one of these statements concerning organic acids is true?

A Phenol is a stronger acid than ethanoic acid.
B Ethanoic acid is a stronger acid than methanoic acid.
C Trinitrophenol is a stronger acid than phenol.
D Phenol is a stronger acid than chloroethanoic acid.

13 A compound has molecular formula C_2H_6O. This compound could be an

A alcohol.
B aldehyde.
C ester.
D alkene.

14 How many isomers can there be of a compound whose molecular formula is C_4H_{10}?

    A   0
    B   2
    C   3
    D   4

15 How many isomers can the molecular formula C_4H_9F represent?

    A   2
    B   4
    C   6
    D   8

16 Analysis revealed a compound to have an empirical formula of CH_3. How many hydrocarbons could have this empirical formula?

    A   an unlimited number
    B   three
    C   one
    D   none

17 A compound consisting only of carbon, hydrogen and oxygen gave the following percentage composition upon analysis: C = 38, H = 12, O = 50. The ratio of atoms present in the compound is $C:H:O = \frac{38}{12}:\frac{12}{1}:\frac{50}{16}$. The empirical formula of the compound is

    A $C_3H_{12}O_3$.
    B C_2H_6O.
    C CH_3O.
    D CH_4O.

18 How many C—H bonds are there in a molecule of butan-2-ol?

    A   10
    B   9
    C   8
    D   7

19 In methanol the C–O–H bond angle is between

    A   180° and 150°.
    B   150° and 120°.
    C   120° and 109°28′.
    D   109°28′ and 90°.

20 Hexane contains the same number of carbon–carbon single bonds as

    A   2,3-dimethylbutane.
    B   3-ethylpentane.
    C   2-methylbutane.
    D   2,2-dimethylpentane.

XVI Aliphatic Hydrocarbons

1 The formula of 2,2,3-trimethylpentane is

  A $(CH_3)_3CCH_2CH_2CH_3$
  B $CH_3C(CH_3)_2CH_2CH(CH_3)CH_3$
  C $(CH_3)_2CHCH(CH_3)CH_2CH_3$
  D $CH_3C(CH_3)_2CH(CH_3)CH_2CH_3$

2 Which one of the following contains a tertiary C—H bond?

  A $C_6H_5CH_3$
  B $CH_3CH=C(CH_3)_2$
  C $C_2H_5CH(CH_3)_2$
  D $CH_3C\equiv CH$

3 Which one of the following will have the lowest boiling point?

  A $CH_3CH_2CH_2CH_2CH_3$
  B $CH_3CH_2CH(CH_3)CH_3$
  C $C(CH_3)_4$
  D $CH_3CH_2CH_2CH_2CH_2CH_3$

4 Which one of the following represents a step in the reaction of chlorine and methane in the presence of ultra-violet light?

  A $CH_4 + Cl^- \rightarrow CH_3Cl + H^-$
  B $CH_4 \rightarrow CH_3^{\cdot} + H^{\cdot}$
  C $CH_4 + Cl^{\cdot} \rightarrow CH_3Cl + H^{\cdot}$
  D $CH_4 + Cl^{\cdot} \rightarrow CH_3^{\cdot} + HCl$

5 How many different monosubstituted chloro derivatives of 2,3-dimethylbutane can be formed?

  A   2
  B   3
  C   4
  D   6

6 Ozonolysis of an alkene (X), followed by hydrolysis yielded a mixture of propanone and methanal. Which one of the following is X?

  A $(CH_3)_2C=C(CH_3)_2$
  B $CH_3CH=CHCH_3$
  C $(CH_3)_2C=CH_2$
  D $C_2H_5(CH_3)C=CH_2$

7 Which one of the following would immediately decolorize bromine dissolved in tetrachloromethane?

A ethane
B benzene
C but-1-ene
D hexane

8 The formula of 3-methylpent-2-ene is

A $CH_3CH=C(CH_3)C_2H_5$
B $C_2H_5CH=C(CH_3)_2$
C $(CH_3)_2C=CHCH_3$
D $(CH_3)_3CCH=CHCH_2CH_3$

9 Which one of the following would be the major product when hydrogen bromide reacts with 2-methylbut-2-ene, $(CH_3)_2C=CHCH_3$?

A 2-bromo-3-methylbutane
B 2-bromo-2-methylbutane
C 1-bromo-2-methylbutane
D 1-bromo-3-methylbutane

10 When ethene reacts with bromine water, the first species formed is

A $C_2H_5^+$
B $C_2H_4OH^-$
C $C_2H_4Br^-$
D $C_2H_4Br^+$

11 Which one of the following compounds is the major product when but-1-ene reacts with concentrated sulphuric acid and the initial product is hydrolyzed?

A butan-1-ol
B butan-2-ol
C butanal
D butanoic acid

12 Of the four carbonium ions shown, the most stable is

A $(CH_3)_3C^+$.
B $CH_3(CH_2)_2CH_2^+$.
C $CH_3CH_2CH^+CH_3$.
D $CH_3CH_2^+$.

13 What volume of hydrogen is required to convert 3 dm³ of propyne (C_3H_4) into propane in the presence of a catalyst—all volumes being measured under the same conditions of temperature and pressure?

A 1 dm³
B 3 dm³
C 6 dm³
D 9 dm³

14 The conversion

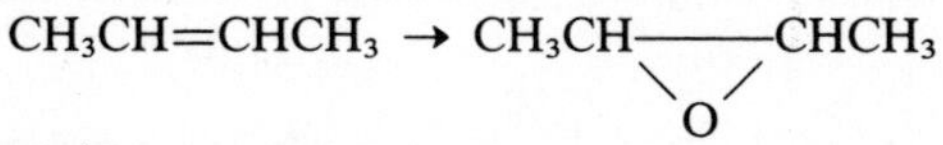

$$CH_3CH{=}CHCH_3 \;\rightarrow\; CH_3CH\overset{\diagdown\;\diagup}{\underset{O}{\rule{1.5cm}{0pt}}}CHCH_3$$

can be brought about by

A ozone followed by hydrolysis.
B concentrated sulphuric acid followed by hydrolysis.
C potassium manganate(VII).
D peroxobenzoic acid (C_6H_5COOOH).

15 Which one of the following would you expect to burn with the sootiest flame?

A methane
B ethane
C ethene
D ethyne

16 A method of preparing methane in the laboratory is to heat the sodium salt of ethanoic acid with soda-lime:

$$CH_3COONa + NaOH \;\rightarrow\; CH_4 + \cdots$$

The other product of the reaction is

A carbon dioxide.
B water.
C sodium carbonate.
D sodium hydrogencarbonate.

17 Ethene can be converted into ethanal by mixing it with air between 20 and 60 °C together with an aqueous solution of copper(II) and palladium(II) chlorides. This is the Wacker–Hoechst process. How many moles of oxygen, O_2, will be required per mole of ethene?

A 0.25
B 0.5
C 1
D 2

18 Ethyne (C_2H_2)

A will *not* react with acidified potassium manganate(VII) solution.
B burns completely to give four moles of products per mole of ethyne.
C forms an explosive solid (Cu_2C_2) when mixed with ammoniacal copper(I) chloride.
D can be made by reducing ethene with hydrogen and a nickel catalyst.

19 What is the molecular formula of the compound that contains one carbon–carbon single bond, one carbon–carbon double bond and one carbon–carbon triple bond?

A C_4H_3
B C_4H_4
C C_4H_5
D C_4H_6

61

XVII Alcohols, Amines, Ethers and Haloalkanes

1 Which one of the following compounds reacts most rapidly with an alcoholic solution of silver nitrate to produce a precipitate?

    A C_4H_9Cl
    B C_4H_9I
    C C_6H_5Cl
    D C_6H_5Br

2 Of the four haloalkanes shown below, the one that is most easily hydrolyzed is

    A 1-bromobutane.
    B 2-bromobutane.
    C 2-bromo-2-methylpropane.
    D 2-chlorobutane.

3 Starting from bromoethane, more than one step is required in the formation of

    A ethanol.
    B methoxyethane.
    C ethyl ethanoate.
    D propanoic acid.

4 The reaction of potassium hydroxide solution with 2-bromopropane results in substitution and elimination reactions occurring. A correct statement about these reactions is that the

    A OH^- ion is acting as a strong base in the substitution reaction.
    B OH^- ion is acting as a nucleophile in the elimination reaction.
    C elimination product is propan-2-ol.
    D elimination reaction is favoured if an alcoholic solution of alkali is employed instead of an aqueous one.

5 An example of a tertiary alcohol is

    A pentan-3-ol.
    B 2-methylpropan-2-ol.
    C propan-2-ol.
    D butan-1-ol.

6 Alcohols

    A react with sodium to form low melting point covalent solids.
    B can act as bases because of the lone pair of electrons on the oxygen atom.
    C have a pH between 7 and 8 in an aqueous solution.
    D have relatively high boiling points because of intra-molecular hydrogen bonding.

7 The first species formed during the dehydration of ethanol by concentrated sulphuric acid is

    A $C_2H_5^+$.

    B $C_2H_5OH_2^+$.

    C $C_2H_5O^-$.

    D $C_2H_5O^+$.

8 The most difficult of the following compounds to oxidize is

    A $CH_3CH_2CH_2OH$.

    B $CH_3CH(OH)CH_3$.

    C $(CH_3)_3COH$.

    D $(CH_3)_2CHCH_2OH$.

9 Oxidation of the alcohol X produced pentan-2-one. X is

    A $CH_3(CH_2)_3CH_2OH$.

    B $(CH_3)_2CHCH_2CH_2OH$.

    C $CH_3(CH_2)_2CH(OH)CH_3$.

    D $CH_3CH_2CH(OH)CH_2CH_3$.

10 Which one of the following alcohols will give a yellow precipitate when warmed with iodine and dilute sodium hydroxide solution?

    A methanol

    B propan-1-ol

    C propan-2-ol

    D 2-methylpropan-2-ol

11 An ester can be formed by the reaction of ethanol with

    A an aldehyde.

    B the sodium salt of a carboxylic acid.

    C an amide.

    D an acid chloride.

12 Which one of the following pairs of compounds will react together under suitable conditions to produce 1-ethoxypropane?

    A CH_3COCl and $CH_3CH_2CH_2OH$

    B C_2H_5ONa and $CH_3CH_2CH_2I$

    C C_2H_5OH and CH_3CH_2COONa

    D C_2H_5ONa and $CH_3CH(I)CH_3$

13 Refluxing ethoxyethane with hydrobromic acid will produce

    A butan-2-ol and bromine.

    B ethanol and bromoethane.

    C ethanal and bromoethane.

    D ethane and 1-bromoethanol.

14 An example of a reaction involving nucleophilic substitution is that between

    A   ethene and hydrogen bromide.
    B   bromoethane and sodium hydroxide solution.
    C   ethanol and concentrated sulphuric acid.
    D   ethanol and potassium manganate(VII) solution.

In questions 15–19 use the responses A–D, shown below, to answer the questions. Each response may be used once, more than once or not at all.

    A RCH_2CH_2OH
    B $RCOOH$
    C $RCH=CH_2$
    D RCH_2COR'

What is produced when

15 RCH_2OH is treated with warm acidified dichromate(VI) solution?

16 RCH_2CH_2Cl is added to a boiling solution of potassium hydroxide in ethanol?

17 RCH_2CH_2OH is warmed with phosphoric(V) acid?

18 RCH_2CHOHR' is treated with warm acidified dichromate(VI) solution?

19 RCH_2CH_2OH is passed over pumice at 400°C?

20 The primary alcohol $C_6H_{11}OH$ is oxidized using acidified dichromate(VI) solution under reflux conditions. The formula of a possible product is

    A $C_6H_{10}O_2$.
    B $C_6H_{12}O_2$.
    C $C_6H_{11}O$.
    D $C_6H_{12}O$.

21 Every ether shares its molecular formula with at least one alcohol. Some also have isomers which are ethers. The latter phenomenon is known as metamerism. Which one of the following has a structural isomer which is also an ether?

    A CH_3OCH_3.
    B $CH_3OC_2H_5$.
    C $C_2H_5OC_3H_7$.
    D $CH_3OCH=CH_2$.

22 An example of a tertiary amine is

    A CH_3NH_2.
    B CH_3NHCH_3.
    C $(C_2H_5)_2NCH_3$.
    D $C_6H_5NHCH_3$.

23 By which one of the following reactions could you prepare a secondary amine?

    A   the Hofmann degradation of ethanamide.

    B   the reductive amination of ethanal.

    C   the reduction of hydrogen cyanide.

    D   the treatment of bromoethane with alcoholic ammonia solution.

24 Which one of the following will have the highest pK_b value?

    A NH_3

    B $C_2H_5NH_2$

    C $(CH_3)_2NH$

    D $(CH_3)_3N$

25 Nitrogen gas can be produced by mixing nitrous acid, $HNO_2(aq)$ at 0 °C with a solution of

    A $C_2H_5NH_2$.

    B $(CH_3)_2NH$.

    C $(CH_3)_3N$.

    D $C_6H_5NH_2$.

XVIII Carboxylic Acids and their Derivatives

1 Ethanoic anhydride, $(CH_3CO)_2O$, can be prepared by reacting

   A   silver ethanoate and chloroethane.
   B   ethanoic acid and concentrated sulphuric acid.
   C   ethanoic acid and ethanol.
   D   sodium ethanoate and ethanoyl chloride.

2 What is the approximate C—C—O bond angle in ethanamide, CH_3CONH_2?

   A   60°
   B   90°
   C   109°28'
   D   120°

3 The formula of 3-methylbutanoic acid is

   A $CH_3CH_2CH(CH_3)COOH$.
   B $CH_3CH(COOH)CH_2CH_3$.
   C $CH_3CH(CH_3)CH_2COOH$.
   D $(CH_3)_3CCH_2COOH$.

4 Which one of the following pairs of compounds will react together to form a secondary alcohol?

   A CH_3COOH and $NaBH_4$
   B C_2H_5COOH and $LiAlH_4$
   C $(CH_3)_2CO$ and $LiAlH_4$
   D $HCOOH$ and concentrated H_2SO_4

5 If one mole of each of the following compounds were separately added to one dm^3 of water in sealed vessels, which would produce the solution with the lowest pH?

   A   ethanoic acid
   B   chloroethanoic acid
   C   phenol
   D   ethanoyl chloride

6 Acid amides $(RCONH_2)$ can be made by the reaction of ammonia with all of the compounds shown below. Which reaction goes via a stable ionic intermediate compound? The reaction between ammonia and

   A   ethanoyl chloride.
   B   propanoic acid.
   C   ethyl propanoate.
   D   ethanoic anhydride.

7 Studies of the formation and hydrolysis of esters involving the labelling of molecules with ^{18}O isotopes gave the following equilibrium:

$$R-\underset{\underset{^{16}O}{\|}}{C}-^{18}O-R'+H_2^{16}O \rightleftharpoons R-\underset{\underset{^{16}O}{\|}}{C}-^{16}O-H+R'-^{18}O-H$$

A true statement about this equilibrium is that

A   the hydrolysis of an ester involves the breaking of a C—O bond.
B   during hydrolysis, the oxygen atom of the water molecule becomes incorporated into the alcohol molecule.
C   the formation of an ester results in the loss of a hydrogen atom by the acid, and the loss of the OH group by the alcohol.
D   the oxygen atom from the carbonyl group of the ester molecule is incorporated into the alcohol during hydrolysis.

8 Ethyl propanoate can be formed by the reaction of

A $CH_3CH_2CH_2OH$ and CH_3COOH.
B $CH_3CH_2CH_2OH$ and CH_3COCl.
C $CH_3CH_2CH_2OH$ and $(CH_3CO)_2O$.
D CH_3CH_2OH and CH_3CH_2COCl.

9 0.2 mole of ethanoyl chloride undergoes complete hydrolysis in water and all the products are dissolved in water. What volume of 2 M sodium hydroxide solution is needed to neutralize the resulting solution?

A $50\,cm^3$
B $100\,cm^3$
C $200\,cm^3$
D $400\,cm^3$

10 Which of the following has the lowest K_a value?

A CH_3COOH
B C_6H_5COOH
C $CH_2ClCOOH$
D CH_2ICOOH

11 The compound $CH_3CONH(C_3H_7)$ can be formed by the reaction of

A   ethanoic anhydride and ethylamine.
B   propanoyl chloride and ethylamine.
C   ethanoyl chloride and propylamine.
D   propanoic anhydride and methylamine.

12 Ethanoyl chloride can be reduced to ethanal by treatment with

A $NaBH_4$.
B Na/C_2H_5OH.
C H_2/poisoned Pd catalyst.
D $LiAlH_4$.

13 The boiling points of the isomeric compounds methyl ethanoate and propanoic acid are 58 °C and 141 °C respectively. This difference in boiling point is best explained by the statement that

    A   propanoic acid has the higher relative molecular mass of the two compounds.
    B   there is hydrogen bonding between propanoic acid molecules.
    C   methyl ethanoate decomposes at 58 °C.
    D   methyl ethanoate molecules form dimers.

14 Propanoic acid can be reduced to propan-1-ol by treatment with

    A   lithium tetrahydridoaluminate(III), (lithium aluminium hydride).
    B   sodium in ethanol.
    C   hydrogen at 200 atm with a copper(II) oxide catalyst.
    D   sodium amalgam in ethanol.

15 When an acid chloride, $RCOCl$, reacts with the sodium salt of a carboxylic acid, $R'COONa$, the product is

    A   the ester $R'COOR$.
    B   the ketone $R'COR$.
    C   the ester $RCOOR'$.
    D   the acid anhydride $(R'CO)O(COR)$.

Using the reaction sequences shown below, answer questions 16–19, which concern the conditions and reactants needed for the various conversions:

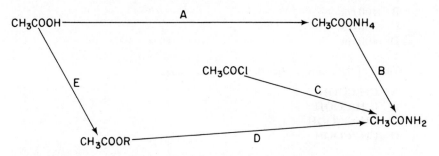

Which step

16 requires the use of a strong acid catalyst?

17 can be brought about by heating alone?

18 results in the production of hydrogen chloride?

19 involves the production of an ionic compound?

XIX Aldehydes and Ketones

1 An example of a planar molecule is

    A   methanol.
    B   methanal.
    C   propanone.
    D   methanoic acid.

2 The formula of phenylethanone is

    A C_6H_5CHO.
    B $C_6H_5CH_2CHO$.
    C $C_6H_5COCH_3$.
    D $C_6H_5COCH_2CH_3$.

3 Which one of the following statements about addition to $C=C$ and $C=O$ bonds is correct?

    A   The $C=C$ bond in ethene is more polar than the $C=O$ bond in methanal.
    B   Addition to $C=C$ bonds generally occurs by nucleophilic attack.
    C   Addition of HCN to a $C=O$ bond proceeds via the protonation of oxygen as a first step.
    D   HBr will add on to $C=C$ bonds in alkenes but HCN will not.

4 The most difficult of the following four compounds to oxidize is

    A   ethanol.
    B   ethanal.
    C   propanone.
    D   benzaldehyde.

5 Which one of the following compounds will give a coloured, crystalline product with 2,4-dinitrophenylhydrazine but will not give a silver mirror with ammoniacal silver nitrate solution?

    A   ethanol
    B   ethanal
    C   benzaldehyde
    D   propanone

6 The reaction between propanone and hydroxylamine (NH_2OH) produces

    A $(CH_3)_2CHNH_2$.
    B $(CH_3)_2CNH_2OH$.
    C $(CH_3)_2C=NOH$.
    D $(CH_3)_2CNH_2$.

7 Triiodomethane (CHI_3) *cannot* be formed by the reaction of an alkaline solution of iodine in potassium iodide and

A ethanal.
B propanal.
C propanone.
D butanone.

8 An example of an addition/elimination reaction is

A $C_6H_5CHO + NaHSO_3 \rightarrow C_6H_5CH(OH)SO_3Na$

B $CH_3CHO \xrightarrow{\text{acidified } MnO_4^-} CH_3COOH$

C $(CH_3)_2CO \xrightarrow{\text{LiAlH}_4 \text{ in ether}} (CH_3)_2CHOH$

D $CH_3CH_2COCH_3 + NH_2OH \rightarrow (CH_3CH_2)(CH_3)C{=}NOH$

9 Lithium tetrahydrioaluminate(III) ($LiAlH_4$) in ether can be used to reduce all four of the compounds shown below. Which one of the reductions would lead to the production of a secondary alcohol?

A $(CH_3CO)_2O$
B $CH_3CH(CH_3)COCl$
C $C_3H_7COCH_3$
D $C_2H_5COOC_2H_5$

10 The ketone R_2CO can be converted into a compound which contains *one more* carbon atom than the original compound by reaction with one *or* more of the following compounds:

  (i)   phenylhydrazine
  (ii)   hydrogen cyanide
  (iii) dilute sodium hydroxide solution
  (iv)   concentrated sodium hydroxide solution

The conversion can be carried out by

A compounds (i) and (ii) only.
B compound (ii) only.
C all the compounds.
D compounds (i), (ii) and (iii).

11 Compounds of general formula $RCH{=}NR'$ are known as Schiff bases. Such a compound can be produced by the reaction of

A propanone and methylamine.
B ethanal and ethylamine.
C ethanal and dimethylamine.
D butanone and propylamine.

70

12 Aldehydes undergo the aldol condensation if treated with dilute sodium hydroxide solution provided that certain conditions are met. The general reaction is:

$$2RCH_2CHO \rightarrow RCH_2CH(OH)CH(R)CHO$$

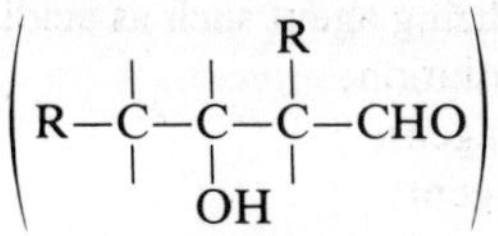

$$\left(\begin{array}{c} R \\ | \\ R-\overset{|}{\underset{|}{C}}-\overset{|}{\underset{|}{C}}-\overset{|}{\underset{|}{C}}-CHO \\ OH \end{array} \right)$$

Which one of the following aldehydes will give the aldol condensation?

A   methanal
B   ethanal
C   2,2-dimethylpropanal
D   benzaldehyde

13 Certain aldehydes can take part in the Cannizzaro reaction which usually occurs in 50% aqueous or alcoholic alkali. In this reaction, benzaldehyde will give rise to

A   phenol and sodium methanoate.
B   phenylmethanol and sodium benzoate.
C   phenylmethanol and sodium methanoate.
D   benzoic acid and sodium methoxide.

14 Aldehydes form condensation products with sodium hydrogensulphite ($NaHSO_3$). These products are usually crystalline. Treating one such product, $CH_3CH_2CH(OH)SO_3Na$, with aqueous alkali will produce

A   propan-1-ol.
B   propanone.
C   propan-2-ol.
D   propanal.

15 2-hydroxypropanoic acid ($CH_3CH(OH)COOH$) can be formed by the reaction of a carbonyl compound with hydrogen cyanide, followed by hydrolysis. Which one of the following compounds would give this reaction?

A   ethanal
B   propanal
C   propanone
D   butanone

16 Ethanol can be dehydrogenated by passing it over a copper catalyst at 300°C. The product will be

A   ethanoic acid.
B   ethene.
C   ethanal.
D   propanone.

71

17 You are given a compound which you are told is either ethanal or propanone. You are told that reaction with each of the following reagents might enable you to identify the compounds:

(i)   a mild oxidizing agent such as acidified dichromate(VI);
(ii) Fehling's solution;
(iii) Tollen's reagent;
(iv) Schiff's reagent;

Which of the reagents would enable you to identify the compounds?

A   none of them
B   all of them
C   (iii) only
D   (i), (ii) and (iii)

18 A compound is analyzed and its molecular formula is found to be $C_6H_{12}O$. It contains an unbranched carbon chain with no alkene or alkyne bonds and possesses one aldehyde functional group. How many possible compounds are there that meet these specifications?

A   none
B   one
C   two
D   more than two

19 If the compound described in question 18 has one branch in the carbon chain, how many possible compounds are there?

A   none
B   one
C   two
D   more than two

XX Aromatic Compounds

1 Benzene

    A   forms two isomeric derivatives of formula $C_6H_4Cl_2$.
    B   undergoes rapid addition with bromine at room temperature.
    C   is an alternative name for the compound cyclohexa-1,3,5-triene.
    D   contains six carbon–carbon bonds of equal length.

2 In benzene

    A   the bonding involves carbon sp^3 hybrid orbitals.
    B   the C—C—C bond angles are all $109°28'$.
    C   the molecular shape is that of a puckered ring.
    D   the C—C bond length is greater than that in ethene.

3 Chlorobenzene can be formed by the reaction of chlorine and benzene in the presence of aluminium chloride. The species which attacks the ring in this reaction is

    A Cl^-.
    B Cl^+.
    C $Cl^{\cdot}$.
    D $[AlCl_4]^-$.

4 When benzene reacts with a mixture of concentrated sulphuric and nitric acids, the species which is formed initially is

5 The following statements refer to methylbenzene and its derivatives. Which one is true?

    A   There are three aromatic isomers of molecular formula C_7H_7Cl.
    B   Methylbenzene will decolorize warm acidified potassium manganate(VII) solution.
    C   Nitration of methylbenzene requires more vigorous conditions than does the nitration of benzene.
    D   The methylbenzene molecule is planar.

6 Which one of the following compounds would react with a solution of sodium nitrite ($NaNO_2$) in dilute hydrochloric acid at 0 °C to form a diazonium compound?

A 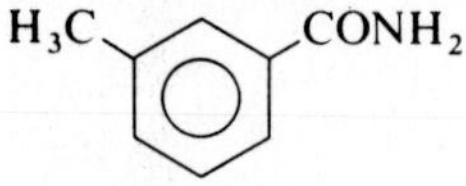   H₃C⟨benzene ring⟩CONH₂

B 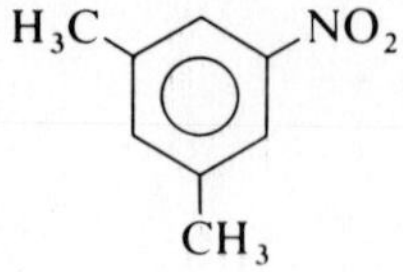   H₃C⟨benzene ring with NO₂ and CH₃⟩

C 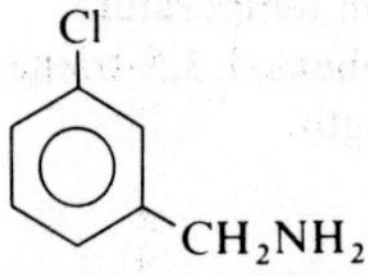   Cl⟨benzene ring⟩CH₂NH₂

D   H₃C⟨benzene ring with NH₂ and Cl⟩

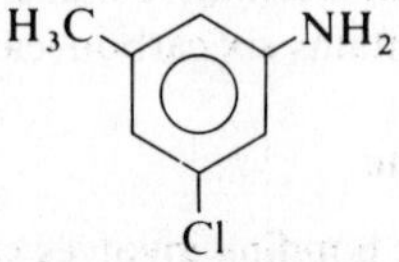

7 An aqueous solution of an organic compound of molecular formula $C_8H_9ClN_2$ decomposed at 25 °C giving bubbles of an unreactive gas and an aromatic compound Z. Another sample of the solution yielded a compound of molecular formula C_8H_9I when it was treated with aqueous potassium iodide solution. The identity of Z is

A   ⟨benzene ring: CH₃ top, H₃C left, Cl right⟩

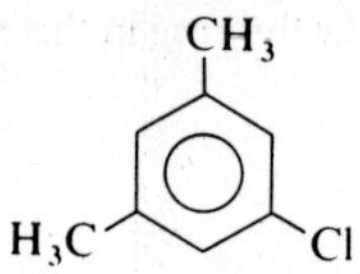

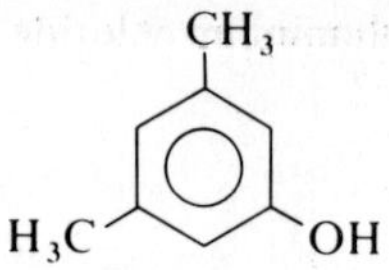

B   ⟨benzene ring: CH₃ top, H₃C left, OH right⟩

C   ⟨benzene ring: CH₃ top, H₃C left, NH₂ right⟩

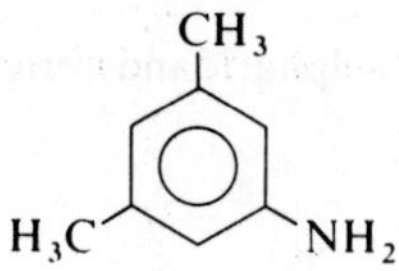

D   ⟨benzene ring: CH₃ top, CH₃ right⟩

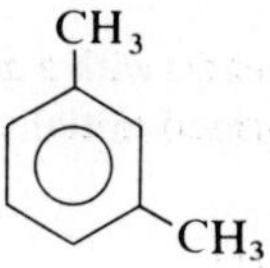

Questions 8 to 10. Use the responses shown below (A–D). Each response may be used once, more than once or not at all.

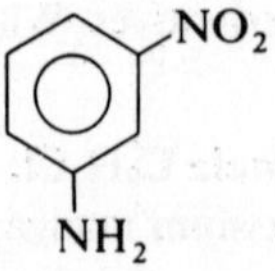

A   ⟨benzene ring⟩NH₂

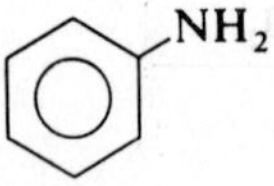

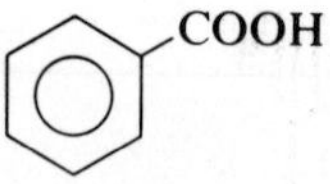

B   ⟨benzene ring⟩COOH

C   ⟨benzene ring with NO₂ and NH₂⟩

D   ⟨benzene ring⟩NO₂

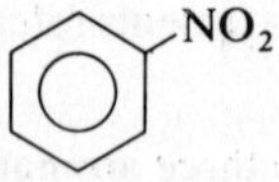

Several compounds are missing from the following reaction sequence. Some of the missing compounds are replaced by question numbers and others by letters. The compounds shown in the responses are those replaced by the numbers:

74

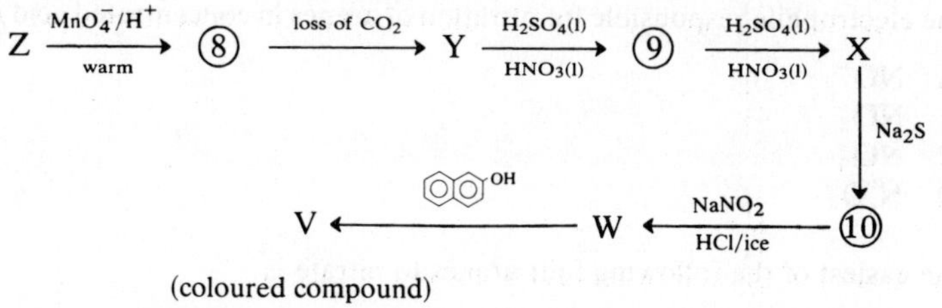

$$Z \xrightarrow[\text{warm}]{MnO_4^-/H^+} \text{⑧} \xrightarrow[\text{loss of } CO_2]{} Y \xrightarrow[HNO_3(l)]{H_2SO_4(l)} \text{⑨} \xrightarrow[HNO_3(l)]{H_2SO_4(l)} X$$

$$X \xrightarrow{Na_2S} \text{⑩}$$

$$V \longleftarrow \text{(naphthol)}^{OH} \longleftarrow W \xleftarrow[HCl/ice]{NaNO_2} \text{⑩}$$

(coloured compound)

Questions 11–16 are concerned with the following multi-stage synthesis in which some of the compounds are represented by their molecular formulae, others by a structural formula and some by letters. The conditions/reactants required are represented by the responses A–E. Use these responses when answering the questions. Each response may be used once, more than once, or not at all.

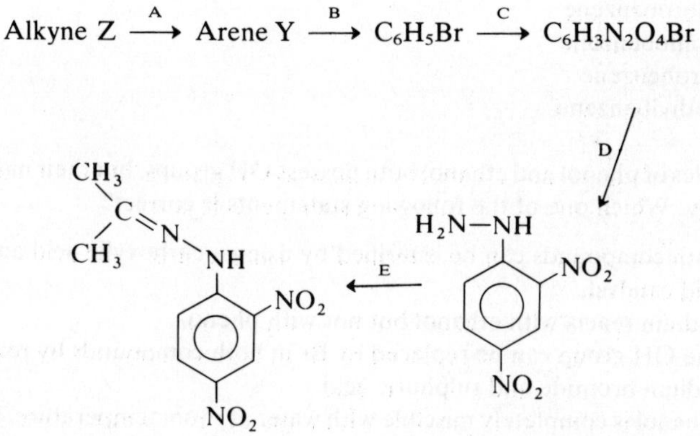

$$\text{Alkyne Z} \xrightarrow{A} \text{Arene Y} \xrightarrow{B} C_6H_5Br \xrightarrow{C} C_6H_3N_2O_4Br$$

$$ \downarrow D $$

structure: $(CH_3)_2C=N-NH-$ attached to benzene ring with NO_2 and NO_2 substituents

$\xleftarrow{E}$ H_2N-NH- attached to benzene ring with NO_2 and NO_2 substituents

Which step

11 requires the use of iron?

12 involves the production of water as a by-product?

13 requires concentrated sulphuric acid?

14 can be brought about by heat alone?

15 is the first one leading to a solid product?

16 might be considered to be polymerization?

17 The electrophile responsible for nitration of arenes in concentrated acid media is

    A NO^+
    B NO_2^+
    C NO_3^+
    D $N_2O_3^+$

18 The easiest of the following four arenes to nitrate is

    A benzene
    B chlorobenzene.
    C nitrobenzene.
    D phenol.

19 Which one of the following would undergo bromination more readily than benzene?

    A chlorobenzene
    B bromobenzene
    C nitrobenzene
    D methylbenzene

20 Molecules of phenol and ethanol both possess OH groups, but their nature differs considerably. Which one of the following statements is correct?

    A Both compounds can be esterified by using a carboxylic acid and a strong acid catalyst.
    B Sodium reacts with ethanol but not with phenol.
    C The OH group can be replaced by Br in both compounds by reaction with sodium bromide and sulphuric acid.
    D Ethanol is completely miscible with water at room temperature, but phenol is not.

21 Compound X, $C_7H_7NO_2$, can be oxidized to a compound of formula, $C_7H_5NO_4$, with acidified manganate(VII) solution. X may be reduced to a compound of formula, C_7H_9N, with tin and concentrated hydrochloric acid. The likely structure of $C_7H_7NO_2$ is

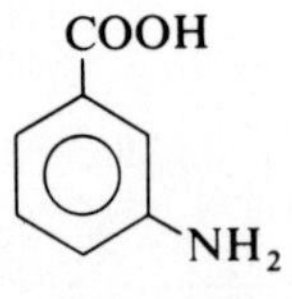

76

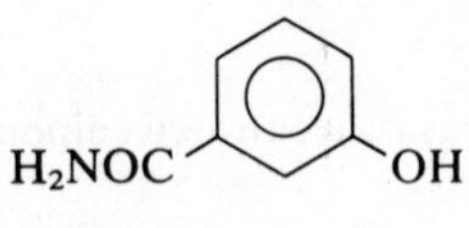

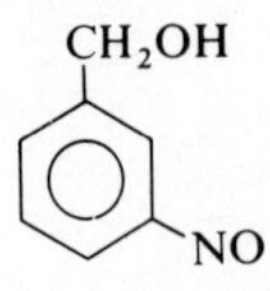

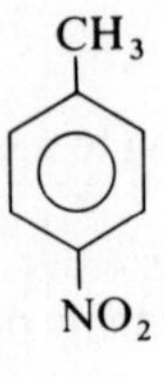

22 In the reaction of benzene and fuming sulphuric acid, the electrophile which attacks the ring is

  A HSO_4^-.
  B SO_3.
  C H^+.
  D SO_3^+.

23 Phenylethanone, 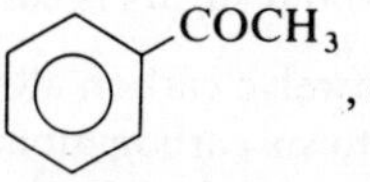 , can be produced by reacting

  A benzene, ethanoyl chloride and aluminium chloride.
  B benzoic acid, ethanol and a strong acid catalyst.
  C benzene, ethanoic anhydride and water.
  D phenol, ethanoyl chloride and sodium hydroxide.

24 Often, in aromatic chemistry, several products can be produced in a substitution reaction. For example, chlorination of benzene can give rise to chlorobenzene, the disubstituted products and then the trisubstituted products. Which one of the following statements about the second substitution products of benzene is true?

  A Formation of 1,2-dimethylbenzene from methylbenzene is more difficult than the formation of methylbenzene from benzene.
  B Formation of 1,3-dinitrobenzene from nitrobenzene is easier than the formation of nitrobenzene from benzene.
  C 1,3-dibromobenzene is formed in preference to 1,2-dibromobenzene in the further bromination of bromobenzene.
  D 1,3-dinitrobenzene is formed in preference to 1,4-dinitrobenzene in the further nitration of nitrobenzene.

25 The side-chain chlorination of methylbenzene

  A does *not* proceed via the homolytic fission of a chlorine molecule.
  B results in the formation of more than one aromatic product.
  C requires a halogen carrier.
  D results in the formation of hydrogen gas as a by-product.

26 How many isomers are there of formula $C_6H_4Cl_2$ which contain a benzene ring?

  A 2
  B 3
  C 4
  D 5

XXI Macromolecules

1 Which one of the following statements about sugars is correct?

A Disaccharides contain more than twelve carbon atoms per molecule.

B Monosaccharides can contain up to six carbon atoms per molecule.

C Disaccharides undergo acidic hydrolysis to yield monosaccharides, but will not undergo enzymic hydrolysis.

D Monosaccharides undergo acidic hydrolysis to yield simpler sugar molecules.

2 Which one of the following classes of monosaccharides contains four carbon atoms, including a ketone group, per molecule?

A ketohexoses

B aldotrioses

C ketotetroses

D aldopentoses

3 Glucose, $C_6H_{12}O_6$,

A is a ketohexose.

B can be hydrolyzed to yield sucrose.

C has two anomers.

D will *not* react with Fehling's solution.

4 Glucose is produced as the *only* product when

A maltose is hydrolyzed with maltase.

B lactose is hydrolyzed with lactase.

C sucrose is hydrolyzed with invertase.

D starch is hydrolyzed with diastase.

5 Natural rubber is a polymer of cis-methylbuta-1,3-diene. The structure of the polymer is:

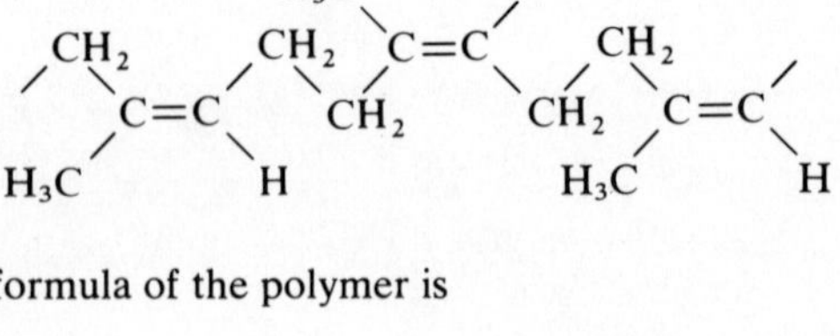

The empirical formula of the polymer is

A C_3H_4.

B C_4H_6.

C C_5H_8.

D C_6H_8.

6 Which one of the following structures represents a block co-polymer?

A   —A—A—A—A—A—A—A—A—A—A—A—A—
B   —A—B—A—B—A—B—A—B—A—B—A—B—
C   —A—B—B—A—A—B—A—B—A—A—B—A—
D   —A—A—A—B—B—B—A—A—A—B—B—B—

7 An isotactic polymer could be formed by the polymerization of

A   ethene.
B   tetrafluoroethene.
C   propene.
D   tetrachloroethene.

8 Which one of the following represents the first propagation step in the free radical polymerization of phenylethene?

$$M^{\cdot} + CH_2{=}CHC_6H_5 \rightarrow$$

A $MCH_2{-}\dot{C}HC_6H_5$
B $CH_2{=}CHM + C_6H_5^{\cdot}$
C $^+CH_2{-}CHC_6H_5 + M^-$
D $MCH_2{-}CH^+C_6H_5 + e^-$

9 Which one of the following represents a section of the polymer nylon 6.6?

A $=N{-}(CH_2)_6{-}N{=}C{-}(CH_2)_6{-}C{=}$ with OH, OH

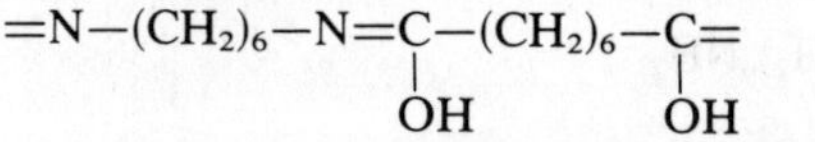

B $-NH{-}(CH_2)_6{-}NH{-}C{-}(CH_2)_6{-}C{-}$ with O, O (double bonds)

C $-NH{-}(CH_2)_5{-}NH{-}C{-}(CH_2)_5{-}C{-}$ with O, O (double bonds)

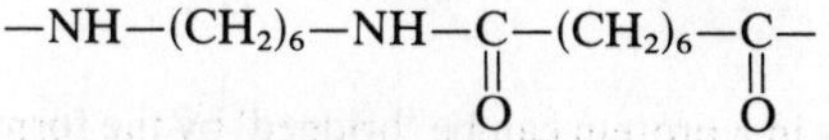

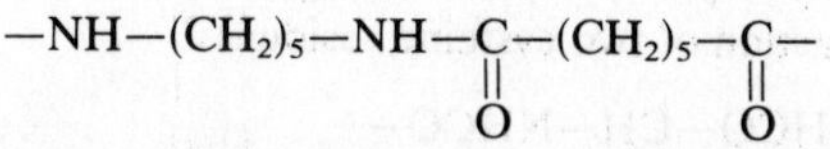

D $-NH{-}(CH_2)_6{-}NH{-}C{-}(CH_2)_4{-}C{-}$ with O, O (double bonds)

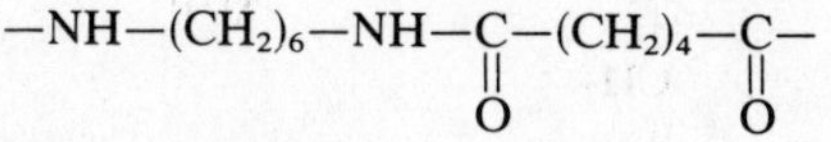

10 Which one of the following does *not* contain peptide links?

A   protein
B   nylon 6
C   terylene
D   methanal-carbamide (formaldehyde-urea) resin

11 Proteins consist of amino acids which have been chemically bonded into chains. How many different chains can be made from the amino acids X, Y and Z if one molecule of each amino acid is used?

A   2
B   4
C   6
D   8

12 Which one of the following pairs of compounds could react together to form a condensation polymer?

A [benzene ring with COCl] + H$_2$N(CH$_2$)$_4$NH$_2$

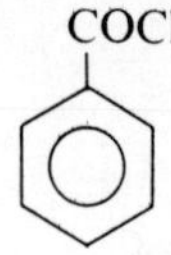

B [benzene ring with COCl top and COCl bottom] + H$_2$N(CH$_2$)$_5$CH$_3$

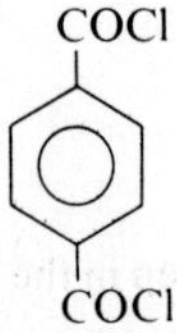

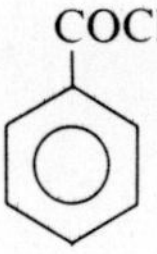

C [benzene ring with COCl] + HOCH$_2$CH$_2$OH

D [benzene ring with COCl top and COCl bottom] + H$_2$N(CH$_2$)$_6$NH$_2$

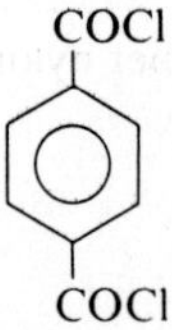

13 Adjacent amino acid chains in a protein can be 'bridged' by the formation of a sulphur–sulphur bond by the reaction of two cysteine residues:

$$-NHCO-CH-NHCO-$$
$$|$$
$$CH_2$$
$$|$$
$$SH$$

The chemical reaction occurring during this bridging is

A oxidation.
B hydrolysis.
C an acid/base reaction.
D reduction.

14 Ethane-1,2-diol, CH$_2$—CH$_2$, can form a condensation polymer with
                                           OH OH

A itself.
B dicarboxylic acids.
C strong mineral acids.
D esters.

Questions 15 to 18 are concerned with the properties of α amino acids. The structure of such acids is usually represented

$$H_2N-CH-COOH$$
$$| \atop CH_2R$$

The properties of an individual amino acid are determined by the identity of R. The responses A to D show two properties of amino acids:

	Nature of the amino acid	R_f value in phenol/water
A	neutral	low
B	acidic	low
C	neutral	high
D	acidic	high

Which properties are those of the amino acid where R is

15 $-C_6H_5$?

16 $-CH_2-COOH$?

17 $-CH(CH_3)_2$?

18 $-OH$?

19 DNA consists of a pair of inter-twined chains. The chains are held together by bonding between adjacent

A base and sugar groups.
B base groups.
C base and phosphate groups.
D sugar groups.

20 The bonding which holds the adjacent chains together in DNA is

A van der Waals bonding.
B covalent bonding.
C hydrogen bonding.
D ion–dipole bonding.

XXII Analysis

1 Which one of the following in aqueous solution will decolorize acidic potassium manganate(VII) solution?

    A Na_2SO_3
    B $Fe_2(SO_4)_3$
    C $NaNO_3$
    D K_2SO_4

2 Which one of the following pairs of substances *cannot* be distinguished by the addition of dilute nitric acid?

    A sodium iodide and sodium nitrite
    B sodium carbonate and sodium sulphate
    C potassium nitrate and potassium nitrite
    D potassium carbonate and potassium hydrogencarbonate

3 A white salt was heated strongly and a colourless gas was evolved. The residue gave brown fumes when treated with dilute hydrochloric acid. The white salt was

    A lead(II) nitrate.
    B sodium hydrogencarbonate
    C potassium nitrate.
    D lead(II) bromide.

4 A colourless solution of a salt, P, gave the following reactions:

    (i) A yellow precipitate was formed with sodium hexanitrocobaltate(III).
    (ii) A cream precipitate was formed with silver nitrate. The precipitate dissolved in excess concentrated ammonia solution but would not dissolve in dilute ammonia solution.

What is the identity of P?

    A sodium iodide
    B potassium bromide
    C ammonium chloride
    D potassium fluoride

5 $25 \, cm^3$ of a solution of sodium chloride and sodium hydroxide was placed in a conical flask and titrated with 0.1 M hydrochloric acid. The solution required $20 \, cm^3$ of acid for neutralization. Potassium chromate(VI) was added and the mixture was then titrated with 0.1 M silver nitrate solution. $30 \, cm^3$ of nitrate solution were required to reach the end-point. What was the molar ratio of sodium chloride to sodium hydroxide in the mixture?

    A $1:2$
    B $1:1$
    C $2:3$
    D $3:2$

6 A student was provided with four colourless solutions, P, Q, R and S. The solutions were dilute hydrochloric acid, barium chloride, sodium carbonate and sodium sulphate, but not necessarily in that order. The student mixed pairs of solutions and the results are shown in the following table. No reaction between the solutions is indicated by –.

	P	Q	R
Q	white ppt.	–	–
R	white ppt.	–	–
S	–	gas evolved	–

What is the correct order of identity of P, Q, R and S?

A Na_2SO_4, Na_2CO_3, $BaCl_2$, HCl
B $BaCl_2$, Na_2CO_3, Na_2SO_4, HCl
C $BaCl_2$, Na_2SO_4, Na_2CO_3, HCl
D Na_2SO_4, HCl, $BaCl_2$, Na_2CO_3

7 A colourless liquid, Z, did not react with ammoniacal silver nitrate solution, but gave a brightly coloured crystalline solid when treated with 2,4-dinitrophenyl-hydrazine solution. Which one of the following is Z?

A C_2H_5CHO
B $C_2H_5OC_2H_5$
C C_2H_5OH
D $C_2H_5COC_2H_5$

8 The following table shows the results of tests upon four organic materials W, X, Y and Z.

	Ammoniacal silver nitrate	2,4-dinitro-phenylhydrazine	Iodoform test
W	silver mirror	coloured ppt.	yellow crystals
X	no reaction	coloured ppt.	yellow crystals
Y	silver mirror	coloured ppt.	no reaction
Z	no reaction	no reaction	yellow crystals

The identities of W, X, Y and Z are

A propanal, butanone, ethanal, propan-2-ol
B ethanal, butanone, propanal, propan-2-ol
C propanal, propanone, butanal, propan-1-ol
D ethanal, propanone, propanal, propan-1-ol

9 Which one of the following compounds will *not* decolorize a warm solution of potassium manganate(VII) to which a few drops of concentrated sulphuric acid have been added?

A but-1-ene
B methylbenzene
C propan-1-ol
D propanone

83

10 In an attempt to distinguish between the three organic liquids methanal, ethanal and benzaldehyde the following test results were obtained. The liquids were labelled R, S and T but not necessarily in that order.

	Addition of sodium hydrogen sulphite solution	Iodoform test	Addition of 2,4-dinitrophenyl-hydrazine solution
R	white ppt.	no reaction	red ppt.
S	no reaction	no reaction	yellow ppt.
T	no reaction	yellow crystals	orange ppt.

The respective letters describing methanal, ethanal and benzaldehyde are

A   R, S, T.
B   S, T, R
C   T, R, S.
D   T, S, R.

11 The reaction between copper(II) sulphate solution and potassium iodide solution is described by:

$$2\ CuSO_4(aq) + 4\ KI(aq) \rightarrow I_2(aq) + K_2SO_4(aq) + 2\ CuI(s)$$

The iodine produced can be quantitatively determined by reaction with sodium thiosulphate(VI) solution:

$$2\ Na_2S_2O_3(aq) + I_2(aq) \rightarrow 2\ NaI(aq) + Na_2S_4O_6(aq)$$

If 25 cm^3 of 0.1 M copper(II) sulphate solution is treated with excess potassium iodide solution and the liberated iodine is determined using 0.1 M sodium thio-sulphate(VI) solution, what volume of thiosulphate(VI) is required?

A   12.5 cm^3
B   25.0 cm^3
C   50.0 cm^3
D   100.0 cm^3

12 The technique used to determine the secondary and tertiary structures of proteins is

A   X-ray diffraction.
B   enzymic hydrolysis.
C   spectroscopy.
D   similar to the method used to determine the primary structure.

13 A mass spectrum of a compound showed prominent lines at mass : charge ratios 15, 16 and 29, together with a small line at ratio 45. The line with ratio 15 could represent the ion fragment

A CH_3^+.
B CH_2^{2+}.
C NH_2^+.
D OH^+.

14 Complete oxidation of 0.01 mole of an ethanedioate required 60 cm^3 of 0.1 M potassium manganate(VII) solution at 70 °C in acidic conditions.

$$16\,H^+(aq) + 2\,MnO_4^-(aq) + 5\,C_2O_4^{2-}(aq) \rightarrow 2\,Mn^{2+}(aq) + 10\,CO_2(g) + 8\,H_2O(l)$$

The formula of the ethanedioate will be

A FeC_2O_4.
B $K_2C_2O_4$.
C $Na_2C_2O_4.2H_2O$.
D $(NH_4)_2C_2O_4$.

15 In the analysis of the primary structure of a protein, the terminal amino group can be made to react with 1-fluoro-2,4-dinitrobenzene (FDNB):

Subsequent to this reaction, the peptide link (—CONH—) is hydrolyzed to produce:

The reagent required for this hydrolysis is

A   aqueous alkali at 60 °C.
B   boiling aqueous alkali.
C   6 M hydrochloric acid.
D   12 M hydrochloric acid.

16 Which one of the following pairs of dilute aqueous solutions would produce a precipitate on mixing?

A $AgNO_3(aq)$ and $KF(aq)$
B $Na_2SO_4(aq)$ and $KI(aq)$
C $BaCl_2(aq)$ and $MgSO_4(aq)$
D $RbCl(aq)$ and $Na_2C_2O_4(aq)$

XXIII Practical Techniques

1 When testing for the presence of a halide ion, dilute nitric acid is added before the addition of silver nitrate solution. The nitric acid is added in order to

    A   prevent the precipitation of silver salts other than the halide.
    B   convert any halide present into the hydrogen halide.
    C   increase the nitrate ion concentration in the solution.
    D   reduce the pH of the solution.

2 Which one of the following is a suitable medium in which to carry out a quantitative oxidation with potassium manganate(VII) solution?

    A $HCl(aq)$
    B $HNO_3(aq)$
    C $H_2S(aq)$
    D $H_3PO_4(aq)$

3 Which one of the following statements about steam distillation is *incorrect*?

    A   The vapour pressure of a mixture of two immiscible liquids is the sum of the vapour pressures of the individual liquids.
    B   A mixture of two immiscible liquids will boil at a temperature below that of either of the pure liquids.
    C   Steam distillation may be used to separate any organic liquid from water.
    D   Steam distillation will take place at a temperature below 100 °C.

4 A moist solution of benzoic acid in ethoxyethane can be dried with

    A   calcium oxide.
    B   sodium hydroxide.
    C   magnesium sulphate.
    D   sodium carbonate.

5 Known masses of a solid, Z, were separately dissolved in water and each made up to 250 cm^3. 25 cm^3 portions of each sample, labelled A, B, C and D, were titrated with silver nitrate solution and the results are shown in the table below:

	Mass of Z used/g (per 250 cm^3 of solution)	Volume of silver nitrate solution used/cm^3
A	0.190	28.5
B	0.182	27.3
C	0.170	25.5
D	0.200	29.5

Which one of the titration results is *not* consistent with the others?

6 Which one of the following indicators would you choose for use in a titration between ethanoic acid and sodium hydroxide solutions?

	Indicator	pH range of the colour change
A	thymol blue	1.2–2.8
B	phenolphthalein	8.2–10.0
C	congo red	3.0–5.0
D	bromocresol purple	5.2–6.8

7 When nitrobenzene is converted into phenylamine by reduction with tin and concentrated hydrochloric acid, the mixture is made alkaline by the addition of aqueous sodium hydroxide solution before steam distillation takes place. Sodium hydroxide solution is added in order to

A dry the phenylamine.
B convert any remaining acid to sodium chloride.
C liberate phenylamine from a complex salt.
D lower the vapour pressure of the mixture.

8 A wet solution of phenylamine in ethoxyethane could be dried with

A aluminium chloride.
B concentrated sulphuric acid.
C phosphorus(III) oxide (P_4O_6).
D calcium oxide.

9 What is the percentage yield in a reaction to prepare the ester $CH_3COOC_2H_5$ ($M_r = 88$) from the acid chloride CH_3COCl if 0.1 mole of acid chloride produces 6.6 g of ester?

A 25
B 33
C 75
D 88

10 In the preparation of propanoyl chloride from propanoic acid, why is sulphur dichloride oxide (thionyl chloride) used in preference to phosphorus pentachloride? Sulphur dichloride oxide

A contains less chlorine atoms per molecule than phosphorus pentachloride.
B produces by-products which are gaseous and easy to remove.
C is easier to handle than phosphorus pentachloride.
D is less likely than phosphorus pentachloride to be hydrolyzed in moist conditions.

11 In the determination of the melting point of a freshly prepared sample of a solid ester, the following apparatus was used:

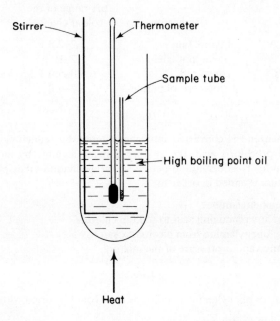

The melting point recorded was 3 °C higher than that recorded in the literature. This could have been because

A   the sample was wet.
B   the rate of heating was too small.
C   the sample contained a small amount of impurity.
D   the rate of heating was too great.

12 One great problem with a multi-step preparation is the considerable loss of material that can occur in the process. How much material could you expect to produce at the end of a three step preparation if you have a 40% yield at each step and you start with 1 mole of material?

$$W \rightarrow X \rightarrow Y \rightarrow Z$$
(1 mole)   (1 mole)

A   0.16 mole
B   0.12 mole
C   0.064 mole
D   0.016 mole

Diagnostic Notes

Contents

I Atomic Structure

1 Answer B *Comprehension of terminology of isotopic symbols*
In the isotopic symbol, m represents the *mass number* (No. of protons + neutrons) and n represents the *atomic number* (No. of protons only). Subtracting the atomic number from the mass number gives the number of neutrons.

2 Answer C *Knowledge of the structure of the atom*
The number of electrons in a neutral atom equals the number of protons. This is numerically equal to the atomic number.

3 Answer D *Comprehension of electronic configurations of atoms*
All elements beyond calcium in the Periodic Table possess d electrons. Only responses **C** and **D** fall into this category, being scandium and iron respectively. The tripositive cation of scandium (response **C**) has the configuration of argon and contains no d electrons.

4 Answer D *Knowledge of electronic configuration of atoms*
The element has the configuration (Argon) $4s^2 3d^5$ and is clearly a transition metal. A transition metal is defined as one which has an incomplete d sub-level of electrons. The element cannot be in groups II, V or VII.

5 Answer C *Comprehension of relationships between atomic and mass numbers*
The atomic number (lower figure) is the number of protons in the nucleus. The mass number (higher figure) is the total number of nucleons (protons and neutrons). The number of neutrons = mass number − atomic number. Hence $^{40}_{20}Ca$ contains 20 protons and 20 neutrons. The other isotopes all contain more neutrons than protons, e.g. argon contains 22 neutrons and 18 protons.

6 Answer C *Knowledge of properties of simple particles and radiation*
The gamma ray, a form of electromagnetic radiation, is not deflected in a magnetic or electric field. The other three particles are charged and will be deflected in an electric field.

7 Answer C *Knowledge of properties of simple particles*
The neutron is uncharged and has a charge/mass ratio of 0. The proton and electron have unit charges; the alpha particle has two positive charges. However, the electron has a minute mass ($\frac{1}{1836}$ of that of the proton), resulting in the largest charge/mass ratio. The mass of an alpha particle is approximately four times that of a proton.

8 Answer B *Application of changes in mass and atomic numbers caused by radioactive decay*
An alpha particle is a charged helium nucleus ($^4_2He^{2+}$) and its loss reduces the mass number by 4 and the atomic number by 2. The beta particle is an electron generated by a neutron disintegrating to form a proton. Its loss does not change the mass number, but causes the atomic number to rise by one. Emission of a gamma ray (electromagnetic radiation) does not change mass or atomic number.

In this change mass number falls by 4 and atomic number changes by $(-2+1) = -1$ to form $^{214}_{83}\text{Bi}$.

9 Answer **A**   *Comprehension of changes in nuclear reactions*
In balanced nuclear equations the sum of mass numbers on each side of the equation must be equal. The sum of atomic numbers on each side of the equation should balance. The particle X should have a mass number of 1 $(14+4=17+1)$ and an atomic number of 1 $(7+2=8+1)$. X is, therefore, a proton: $^{14}_{7}\text{N} + ^{4}_{2}\text{He} \rightarrow ^{17}_{8}\text{O} + ^{1}_{1}\text{p}$.

                           Application of knowledge of changes during
10 Answer **A**   *radioactive decay*
In this process the mass number falls by sixteen and the atomic number by seven. The mass number is changed only by alpha particle emission and so four alpha particles are emitted $(4 \times 4 = 16)$. However, loss of four alpha particles would reduce the atomic number by eight. Loss of one beta particle in addition will reduce this loss to seven.

$$^{228}_{90}\text{Th} \rightarrow ^{212}_{83}\text{Bi} + 4\,^{4}_{2}\text{He} + ^{0}_{-1}\text{e}$$

This change would take place by way of five separate emissions.

11 Answer **C**   *Knowledge of properties of simple radiations*
The statements concerning alpha particles and gamma rays are true. Gamma rays are a form of electromagnetic radiation emitted at the velocity of light by a nucleus, in an excited state. A beta particle, however, is emitted by the nuclear process:

$$\text{neutron} \rightarrow \text{proton} + \text{electron}$$

and is *not* a valency electron.

12 Answer **C**   *Comprehension of rates of radioactive decay*
The intensity of radiation from a source falls to one half of its original value during one half-life. During the next half-life it falls to $\frac{1}{2} \times \frac{1}{2} = \frac{1}{4}$ of its original intensity. After three half-lives have passed it reaches $\frac{1}{2} \times \frac{1}{2} \times \frac{1}{2} = \frac{1}{8}$ of its original intensity. 48 days represent three half-lives, and so one half-life is 16 days.

13 Answer **B**   *Application of frequency relationships in the hydrogen spectrum*
Substituting the values for n_1 and n_2 into the equation, we find:

A Frequency $= \dfrac{3Rc}{4}$ B Frequency $= \dfrac{8Rc}{9}$

C Frequency $= \dfrac{5Rc}{36}$ D Frequency $= \dfrac{21Rc}{100}$

The energy levels converge as the value of n increases. Transitions to and from energy level 1 involve the highest energy changes and highest frequencies. The transition $(n=3) \rightarrow (n=1)$ involves a larger energy change than $(n=2) \rightarrow (n=1)$, and therefore has a greater frequency of radiation associated with it.

14 Answer **B**   *Knowledge of Lyman series of transitions*
The Lyman series involves electron transitions to energy level 1. The first line in
the series is caused by transition $(n = 2) \rightarrow (n = 1)$, the second line results from
transition $(n = 3) \rightarrow (n = 1)$.

15 Answer **B**   *Knowledge of nuclear fission processes*
The reaction involves the breaking up (fission) of uranium-235 into two smaller
nuclides. Two neutrons are produced for every neutron which is captured by an
atom of uranium-235. The process is therefore self-propagating and is called a
chain reaction. It can be controlled by absorbing the neutrons produced. The
process is accompanied by the release of large amounts of energy owing to the
mass loss.

16 Answer **C**   *Comprehension of the arrangement of electrons within orbitals*
The 3d sub-level contains five orbitals, each capable of holding two paired
electrons. The only possible arrangements are:

$$(\uparrow\downarrow)(\uparrow\downarrow)(\uparrow)(\uparrow)(\uparrow)$$

and

$$(\uparrow\downarrow)(\uparrow\downarrow)(\uparrow\downarrow)(\uparrow)(\;\;).$$

The first gives rise to three unpaired electrons and the second to only one. Many
transition metal ions can exhibit two possible electron arrangements as high-
spin and low-spin. It is the ligands present in the complex which determine the
actual electron arrangement.

17 Answer **A**   *Application of knowledge of electron configuration*
From the electronic configurations given it can be seen that sodium is the only
element here with an unpaired electron.
Na: $1s^2\, 2s^2\, 2p^6\, 3s^1$
Mg: $1s^2\, 2s^2\, 2p^6\, 3s^2$
Ar: $1s^2\, 2s^2\, 2p^6\, 3s^2\, 3p^6$
Zn: (Argon) $4s^2\, 3d^{10}$

II Structure and Bonding

1 Answer **B**   *Knowledge of crystal lattices*
The sodium chloride lattice is considered to be composed of two face-centred cubic lattices which penetrate each other, one lattice being composed solely of sodium ions and the other solely of chloride ions. The co-ordination number of each ion in this case is six. That is to say each ion of one kind is surrounded by six ions of the other kind within the lattice. Caesium chloride consists of two inter-penetrating simple cubic lattices, one lattice being composed of caesium ions and the other of chloride ions. Here, each ion is surrounded by eight of the other ions within the lattice.

2 Answer **C**   *Application of the Bragg equation*
The Bragg equation tells us that $n\lambda = 2d \sin \theta$, where n is the number of wavelengths by which the diffracted ray is out of phase with that reflected from the upper layer. If $n = 2$, then $2\lambda = 2d \sin \theta$ and $\lambda = d \sin \theta$.

3 Answer **B**   *Knowledge of conduction processes*
An electric current is effected by the movement of electrons or ions. Graphite possesses 'delocalized' electrons which are able to move through the structure. An ionic substance in the molten state or in solution is able to conduct a current because the ions are free to move through the bulk of the material. In the solid state, however, ions are in a fixed lattice and consequently cannot conduct electricity.

4 Answer **D**
Comprehension of the principles determining the degree of ionic character in a compound
Most bonds lie between being perfectly ionic or covalent because the linked atoms have different electronegativities. The electronegativity of an atom is a measure of the power of the atom to attract electrons to itself within a bond. The ionic character of a compound is smallest where the two atoms involved have similar electronegativities. Aluminium has the highest electronegativity value of the four metals because it is the furthest to the right in the Periodic Table. Therefore, $AlCl_3$ has the smallest degree of ionic character of the four chlorides.

5 Answer **D**   *Comprehension of factors determining bond angles in molecules*
The bond angle in a tetrahedral molecule, e.g. SiH_4, is 109°28'. A carbon dioxide molecule is linear (bond angles 180°), whilst boron trifluoride is an example of a trigonal planar molecule (bond angles 120°). Phosphorus pentafluoride is an example of a trigonal bipyramidal arrangement and possesses bond angles of 90°, 120° and 180°.

6 Answer **C**   *Comprehension of bonding in simple molecules/ions*
The species H_3O^+, H_2S, NH_3 and BF_3 all involve covalent bonding. The ion H_3O^+ also involves dative bonding. There are two lone pairs on the sulphur atom in H_2S. Similarly the oxygen atom in water (H_2O) possesses two lone pairs, but one is used in the dative bond to H^+ in H_3O^+. The nitrogen atom in NH_3

possesses a lone pair but the outer electrons on the boron atom $(2s^2, 2p^1)$ are all used in bonding in BF_3. The $1s^2$ electrons in the boron atom do not constitute a lone pair because they cannot be involved in bonding.

7 Answer **A**   *Comprehension of principles determining shape of molecules*
The shapes of simple molecules may be predicted from the number of lone and bonding pairs possessed. Six pairs result in an octahedral shape, e.g. SF_6, whilst three bonding pairs around a central atom with one lone pair will produce a trigonal pyramidal shape, e.g. NH_3. The water molecule with two lone pairs on oxygen and two bonding pairs is planar and V-shaped. A molecule possessing three bonding pairs only will be trigonal planar in arrangement, e.g. BF_3.

8 Answer **D**   *Knowledge of dipoles in molecules*
A dipole is set up by uneven arrangement of bonding electron density between two atoms (usually of different electronegativities), e.g. H—F, H—O (in H_2O), H—N (in NH_3). Although each C—Cl bond in CCl_4 is polar, the four bonds are arranged symmetrically in space and effectively cancel out each other's polarity.

9 Answer **B**   *Knowledge of causes of hydrogen bonding*
Hydrogen bonding can occur between molecules possessing a dipole and a hydrogen atom. The strength of the dipole, and hence the amount of hydrogen bonding, depends on the difference in electronegativity between hydrogen and the other atom. In these examples oxygen is the most electronegative atom and therefore hydrogen bonding is most prevalent in water.

10 Answer **D**   *Application of knowledge of bonding in ionic compounds*
The term isoelectronic means possessing the same number of electrons. In calcium chloride both the Ca^{2+} and Cl^- ions possess 18 electrons.

11 Answer **A**   *Comprehension of bonding in molecules and ions*
The dimeric Al_2Cl_6 molecule is formed by two $AlCl_3$ molecules linked by two Cl—Al co-ordinate linkages. The complex ion $[Fe(H_2O)_6]^{3+}$ is a typical co-ordination complex, with six water molecules bonded to the central ion. The C_2H_5OH molecule is totally covalently bonded, whilst a co-ordinate N—H linkage exists in NH_4Cl.

12 Answer **B**   *Knowledge of bonding in various states of water*
Hydrogen bonding is most important in determining the properties of water and ice. Above its boiling point most, but not all, water exists as single molecules. In ice each oxygen atom is surrounded by four hydrogen atoms (two bonded covalently and two by hydrogen bonding). The O—H hydrogen bond is much longer than the covalent O—H bond resulting in an open, but rigid structure.

13 Answer **B**   *Knowledge of dipoles in molecules*
A polar liquid will be attracted to a charged object by static electricity. Water, propanone and trichloromethane all possess a dipole and will be attracted. A hexane molecule is non-polar as carbon and hydrogen are very similar in electronegativity.

14 Answer **B**   *Knowledge of X-ray diffraction*
The extent to which atoms diffract X-rays is dependent upon the number of electrons possessed by the atom. An atom such as hydrogen with only one electron cannot diffract X-rays to any appreciable extent. Consequently

hydrogen atoms in a molecule are not detected when the molecule is subjected to X-ray diffraction.

15 Answer **D** *Knowledge of close-packed structures*

Close-packed structures (cubic close-packed and hexagonal close-packed) have a co-ordination number of 12 and because of their close-packed nature give rise to structures with generally higher densities than those having a body-centred cubic structure such as the alkali metals. In close-packed structures empty space within the cube is minimized by the closest possible approach of the constituents.

16 Answer **A** *Comprehension of conditions necessary for isotropic crystals*

Isotropic crystals do not, amongst other things, rotate a plane of polarized light because they have spherical constituents in the crystals. Any crystal containing non-spherical ions or molecules such as NO_3^-, CO_3^{2-} (planar) or C_2^{2-} (rod-like) will render the crystal asymmetrical and hence anisotropic.

17 Answer **C** *Comprehension of molecular architecture*

The shapes of the molecules are as shown below, lone pairs are indicated by xx.

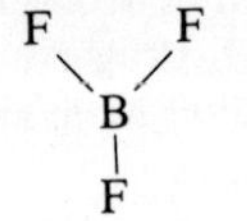

F   F
 \ /
  B
  |
  F

The bond angles are 120° and there is a symmetrical distribution of the fluorine atoms about the boron atom.

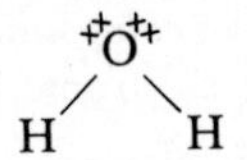

$^{xx}O^{xx}$
 / \
H   H

A V-shaped molecule which has a bond angle of approximately 104°.

Cl—Be—Cl

The linear molecule results from the use of the $2s^2$ electrons in bonding. They get as far apart as possible when forming bonds.

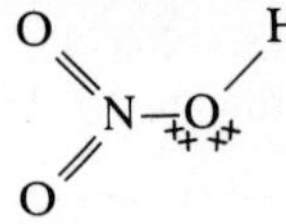

O   H
 \\   /
  N—O
 // $_{xx}$
O

The approximate O—N—O bond angle is 120° and the N—O—H bond angle is approximately 109° 28′.

18 Answer **B** *Comprehension of bond length differences between single, double and triple bonds*

The structures are as follows:

$$N\equiv N \qquad H_2N—NH_2 \qquad H_3C—N=N—CH_3$$

Single bonds are longer than double bonds which are longer than triple bonds. This arises because of the greater electron density in the multiple bonds.

19 Answer **B** *Application of knowledge of bond structures to an unfamiliar dot and cross diagram*

20 Answer **D**

21 Answer **B**

22 Answer **A**

23 Answer **B**

24 Answer **B**

25 Answer **A**

Bonds p, r, s and t are single bonds, which are thought of as containing one electron from each constituent atom. Bond q is a double bond with two electrons regarded as coming from each constituent atom. Sulphur has two lone pairs of electrons and nitrogen one.

III States of Matter

1 Answer A *Application of Avogadro's law*

Equal volumes of gases at the same temperature and pressure contain equal numbers of molecules and, therefore, equal numbers of moles. Thus,

$$\text{No. of moles} = \frac{0.23}{M_r(Y)} = \frac{0.29}{58}$$

and the relative molecular mass of $Y = \dfrac{0.23 \times 58}{0.29}$

2 Answer B *Knowledge of the behaviour of ideal gases*

Predictions of the behaviour of ideal gases assume that there is negligible intermolecular attraction and that the molecules occupy negligible volume. These assumptions are incorrect, particularly at high pressure and low temperature when molecules are close to each other and moving slowly.

3 Answer D *Comprehension of Graham's law of diffusion*

Graham's law states:

$$\text{rate of diffusion} \propto \sqrt{\frac{1}{\text{density}}}$$

$$\text{Density at s.t.p. for any gas} = \frac{\text{molar mass}}{22.4} \text{ g dm}^{-3}$$

Therefore, rate of diffusion $\propto \sqrt{\dfrac{1}{\text{molar mass}}}$

It follows that:

$$\frac{\text{rate of diffusion of H}_2\text{(g)}}{\text{rate of diffusion of O}_2\text{(g)}} = \sqrt{\frac{\text{molar mass of oxygen}}{\text{molar mass of hydrogen}}} = \sqrt{\frac{32}{2}} = \frac{4}{1}.$$

4 Answer C *Application of knowledge of the ideal gas equation*

The ideal gas equation can be written

$$\frac{p_1 V_1}{T_1} = \frac{p_2 V_2}{T_2}$$

for a fixed mass of gas. When one set of data refers to s.t.p. it becomes:

$$\frac{p_1 V_1}{T_1} = \frac{760 V_2}{273} \quad \text{and so} \quad V_2 = \frac{273 \, p_1 V_1}{760 \, T_1}.$$

5 **Answer C** *Comprehension of the distribution of molecular velocities in a gas*
Raising the temperature of a sample of gas causes a shift to higher velocities. The peak is flattened showing a wider distribution of velocities. The lowering of the peak shows that fewer molecules have the most probable velocity at a higher temperature. T_B is the higher of the two temperatures.

6 **Answer D** *Knowledge of effect of variables on vapour pressure*
If water is placed in a sealed vessel it will reach a situation of dynamic equilibrium with the vapour above it. The equilibrium position and therefore the vapour pressure will change with temperature. If there is addition of water or an increase in the volume of the vessel the vapour pressure will not change. Addition of a non-volatile solute lowers the vapour pressure.

7 **Answer A** *Comprehension of partial pressures of gas mixtures*
If 1 mole of $N_2O_4(g)$ becomes 50% dissociated, 0.5 mole of $N_2O_4(g)$ remains *and* 1 mole of $NO_2(g)$ is produced. The ratio of moles, which is equal to the ratio of partial pressures $= 0.5 : 1 = 1 : 2$.

8 **Answer B** *Comprehension of average kinetic energy of gases*
At the same temperature the average kinetic energy of all gases is constant. For one mole of gas the average kinetic energy $= \frac{1}{2} M_r c^2$ and so the gas with the smallest value of M_r has the largest value of c^2. (M_r values: $NH_3 = 17$, $CO = 28$, $O_2 = 32$, $H_2S = 34$.)

9 **Answer C** *Comprehension of Avogadro's law*
From Avogadro's law it follows that the volume of a gas at s.t.p. is proportional to the number of moles of gas present. The number of moles represented by 1 g of the gases listed are:

A $\frac{1}{28}$ B $\frac{1}{18}$ C $\frac{1}{16}$ D $\frac{1}{30}$.

The greatest volume of gas will be that represented by the largest number of moles i.e. C.

10 **Answer B** *Application of knowledge of the ideal gas equation*
The ideal gas equation states that $pV = nRT$, so that the gas constant, $R = pV/nT$. Consequently, it must have units: $(\text{Pressure})^1 (\text{Volume})^1 (\text{Moles})^{-1} (\text{Temperature})^{-1}$. The only response which fulfils these conditions is **B**. The gas constant R is usually quoted in $J\ K^{-1}\ mol^{-1}$.

11 **Answer A** *Application of knowledge of the ideal gas equation*
It is possible to replace V and n in the ideal gas equation by m/d and m/M respectively. Using these substitutions, it can be seen that response **A** is a valid expression,

$$pV = \frac{RTm}{M} \quad \text{and so} \quad T = \frac{pMV}{mR}.$$

Application of Avogadro's law to volume changes during a reaction
12 **Answer A**
A decrease in volume of one third occurs where the stoichiometric equation shows that a reaction going to completion will produce a decrease of one third in the number of moles of gas present. This is the case in response **A** where three moles of reactants produce two moles of products.

13 Answer **C**   *Comprehension of molar volumes of gases*
One mole of a gas at s.t.p. occupies approximately 22.4 dm^3. 16 g of O(g) represents one mole of gaseous oxygen *atoms* and will therefore occupy approximately 22.4 dm^3 at s.t.p. Responses **A** and **D** both represent 0.5 mole of gaseous species, whilst response **C** represents 0.33 mole.

                                    Comprehension of conditions required for accurate
14 Answer **D**   *experimental determination of M_r of a volatile liquid*
It is essential that the boiling point of the liquid be lower than the temperature of the syringe so that rapid, complete vaporization takes place. The syringe does not need to be empty, all that is necessary is that the initial and final gas volumes in the syringe are known. The heat of vaporization of the liquid is irrelevant. The liquid under test must be stable at the temperature of the syringe, otherwise decomposition may occur.

IV Periodicity

1 Answer **D**   *Knowledge of ionization processes*
The ionization energy is the enthalpy change when one mole of electrons is removed from one mole of atoms/ions in the gas phase, producing one mole of gaseous ions. The first ionization energy is the enthalpy change on converting the element to the monopositive ion. The second ionization energy is the enthalpy change on converting the monopositive ion to the dipositive ion.

2 Answer **C**   *Knowledge of trends within a group of elements*
Electropositivity, atomic radius and ionic radius generally increase down a group of elements. However, as the atomic radius increases the outer electrons are further from the nucleus and the energy required to remove them is reduced. The outer electrons are shielded from the full attractive force of the nucleus by the presence of complete inner levels of electrons and so the first ionization energy decreases.

3 Answer **A**   *Knowledge of trends in heat of fusion of elements*
The heat of fusion is the energy absorbed when one mole of atoms of a solid element is converted to liquid at its melting point. Group IV elements (atomic numbers 6, 14, 32) have very strong interatomic forces in the solid state. They have the highest values of heat of fusion on the graph because large quantities of energy are required to overcome these forces before the change of phase is complete.

4 Answer **B**   *Knowledge of electronic configurations of transition metals*
The addition of three more electrons to the (Argon) $4s^2 3d^1$ configuration of scandium would at first seem to lead to a $4s^2 3d^4$ configuration. However, a significant stabilization is achieved by half filling the 3d subshell by transfer of 1 of the 4s electrons. Similarly the configuration of copper is (Argon) $4s^1 3d^{10}$ instead of of the expected $4s^2 3d^9$.

                              Knowledge of the variation of atomic radius within
5 Answer **A**   *a group of the Periodic Table*
As a group is descended the nuclear charge and the number of electrons in the atom are both increasing. These two factors, however, do not cancel out. The electrons are being added to orbitals which are shielded from the full attractive effects of the extra protons in the nucleus, by those electrons already present in the inner orbitals. Consequently the distance of the outer electrons from the nucleus increases.

6 Answer **B**   *Knowledge of trends in acidity of oxides*
The metallic character of elements decreases on crossing a period from left to right. Consequently there is an increase in acidity in the oxides and Cl_2O_7 is the most acidic of the listed oxides.

7 Answer **C**   *Knowledge of factors controlling properties of elements*
The anomalous properties of a 'head element' are best explained by its small

atomic radius compared with other members of the group, and by its greater electronegativity. The 'head element' of the group has the largest first ionization energy. There is no consistent pattern linking the melting point of an element with its position in a group.

8 Answer **D**   *Comprehension of trends in atomic radius in the Periodic Table*
Atomic radius increases down a group of elements as more orbitals are filled. (i.e. $F < Cl < Br$). Across a period there is a slight decrease in radius to the halogen. As an extra proton and electron are added the shielding of the outer electrons by the inner electrons becomes less effective. The radius increases as the effective nuclear attraction decreases (i.e. $N > O > F$). Therefore $O > F$ and $S > O$.

9 Answer **D**   *Comprehension of changes in radius when ions are formed*
The radius of a negative ion is greater than that of its parent atom. It has more electrons than protons and so the effective attractive force on outer electrons is less than in the atom (i.e. $Cl^- > Cl$ and $Si^{4-} > Si$). The proton : electron ratio in a positive ion is greater than in the neutral atom because of the loss of electron(s) in the formation of the ion. Each electron in the positive ion experiences an average force of attraction upon it which is greater than that in the neutral atom and consequently the radius is smaller ($Na^+ < Na$).

10 Answer **C**   *Comprehension of properties of metals and non-metals*
The statement is supported by **A, B** and **D** in that, in general, metallic elements conduct electricity well, whereas non-metallic elements do not; metallic elements are generally easily attacked by dilute acids but non-metallic elements require oxidizing conditions. Semi-metals are susceptible to attack by both acids and alkalis; metallic elements form simple ions by electron loss in their lower oxidation states whereas non-metallic elements usually form ions through covalent combination with oxygen, e.g. CO_3^{2-}, $C_2O_4^{2-}$, SiO_4^{4-}, etc. The existence of allotropy (elemental polymorphism) is by no means confined to non-metals, although some of the elements with well known allotropic forms are non-metallic e.g. sulphur, phosphorus and carbon. The existence of allotropic forms in metals is more widespread than generally thought, e.g. tin.

11 Answer **B**   *Comprehension of trends in ionization energies*
The maximum value of first ionization energies along a period is shown by the noble gas. It is difficult to remove an electron from a stable, full energy level. The minimum value along the period is shown by the Group I element. The fall in value from 2080 to 494 kJ mol^{-1} represents the change from a noble gas (Ne) to a Group I element (Na).

12 Answer **C**   *Comprehension of diagonal relationships between elements*
A diagonal relationship results from similarities in electronegativity and radius between two elements. This causes similarity in some chemical properties. Li and Mg have similar electronegativities and radii. On going across the period Li to Be there is a slight increase in electronegativity which is then cancelled by moving down from Be to Mg.

13 Answer **C**   *Application of knowledge of trends within a group of elements*
Elements in Group VI have an electronic arrangement ending in $ns^2 np^4$ and they form oxides of formula XO_2 and XO_3 (cf. sulphur). They follow the general

trends concerning atomic radius and ionization energy, and consequently the first ionization energy of tellurium will be less than that of sulphur.

14 **Answer D**   *Application of knowledge of electronic structure*
All three species have the noble gas configuration of argon. However, although they are isoelectronic, they do not contain the same number of protons and neutrons. Their nuclear charges are in the order $Ca^{2+} > K^+ > Ar$ and the effective nuclear charge experienced by the outer electrons increases in the same order.

15 **Answer D**   *Application of knowledge of electronic arrangements*
Manganese, potassium and scandium all have an incomplete third energy level together with an unfilled fourth energy level and consequently qualify under this definition as transition elements. Zinc, however, has the third level completed with an arrangement (Argon) $3d^{10} 4s^2$, which includes ten electrons in the 3d sub-level. Clearly this is not a satisfactory definition of a transition element because it encompasses potassium and calcium. (See question 16)

16 **Answer D**   *Application of knowledge of electronic configurations*
Sc^+ and Sc^{3+} contain no d electrons, their configurations being (Argon) $4s^2$ and (Argon) respectively. Zn^{2+} has a full 3d sub-level. Only V^{3+} with its two 3d electrons can be regarded as being a transition metal under this definition.

17 **Answer B**   *Application of knowledge of the Periodic Table*
The atomic numbers of the elements concerned are $H = 1$, $Li = 3$, $C = 6$, $N = 7$, $F = 9$, $Na = 11$ and $Mg = 12$. Consequently, only the species in response **B** contain equal numbers of electrons (14).

18 **Answer D**   *Application of information given about ionization energies*
Group IV elements must have four electrons removed before a noble gas core is revealed. Thus, a relatively large difference is expected between the fourth and fifth ionization energies. This is encountered only in **D**. **A** is an element which possesses *at least* five electrons outside its noble gas core, **B** is a Group I metal and **C** is a Group III element.

19 **Answer B**   *Application of knowledge of the Periodic Table*
Non-ionic binary compounds between adjacent elements in the same period, e.g. NO, *must* contain an odd number of electrons because one of the elements will have an odd atomic number and the other an even one. The other situations cited might give rise to an odd number of electrons, e.g. NO_2, but the examples are exceptional.

20 **Answer B**   *Knowledge of the inert pair effect*
This is an example of the inert pair effect where a pair of outer level electrons exist which are not used in bonding. The example given here is thallium ((Xenon) $6s^2 5d^{10} 6p^1$). The metal shows oxidation states +I and +III. Other examples are bismuth and lead. This should not be confused with the phenomenon of the lone pair. Elements E and H, although transition metals, have oxidation states which exclude them from the conditions stated in the question.

21 Answer **A**
The most non-metallic element capable of forming an oxide is **A**. This will produce the most acidic oxide.

22 Answer **C**   *Application of knowledge of the Periodic Table*
This is an example of the diagonal relationship where the first member of a group is more closely related in its properties to the second member of the next group than to the corresponding element in its own group. Good examples are Li and Mg; Be and Al; B and Si.

V Thermodynamics

1 Answer B *Application of knowledge to the calculation of enthalpy changes*
The enthalpy change for the process is determined by multiplying the 'mass of liquid' by the 'specific heat capacity' of the liquid and by the temperature change. This is $100 \times 4.2 \times 10 = 4.2 \times 10^3$ J. This refers to 0.1 mole of material and so the enthalpy change is $4.2 \times 10^3 \times 10$ J mol^{-1} or 42 kJ mol^{-1}.

2 Answer B *Knowledge of thermodynamic terminology*
Enthalpy changes are measured under conditions of constant pressure and internal energy changes under conditions of constant volume. Enthalpy changes refer to the energy released as *heat* from reactions whereas free energy is the amount of energy released which is available to perform useful work.

3 Answer D *Comprehension of bond energies*
The bond energy of the O—H bond is the amount of energy required to break the bond, giving rise to its constituent atoms. Bond energies are endothermic; the enthalpy changes of the reactions shown are all exothermic, but that of response **D** will be numerically equal to the bond energy although opposite in sign. Response **C** represents an ionic reaction and thus involves more than simple O—H bond formation. In response **B** there are two O—H bonds and an O—O bond being formed as well as H—H and O=O bonds being broken. Similar considerations disqualify response **A**.

4 Answer A *Knowledge of bond energies*
Multiple bonds (C≡C, C=C, C=O, etc) are stronger than their corresponding single bonds because of the increased electron density between the atoms. Thus, more energy has to be used to break the bonds. On these simple considerations, the C≡C bond will be the one with the largest bond energy. In reality, however, the C≡C bond energy of ethyne is 837 kJ mol^{-1}, whilst that of the C=O bond in carbon dioxide is only slightly smaller at 803 kJ mol^{-1}. The C=O bond energy in aldehydes and ketones is lower than that found in carbon dioxide.

Knowledge of conditions necessary for the measurement
5 Answer B *of standard enthalpy changes*
Although standard conditions refer to 760 mmHg (1 atm pressure) and 298 K, reactions can be allowed to take place under other conditions (sometimes they have to be so that the reactions will occur at all) and corrections made to the values obtained. The reactants and products, however, must be pure.

Application of knowledge of the enthalpy changes associated
6 Answer D *with electron loss and gain*
The ionization of atoms is an endothermic process as electrons are being withdrawn against the attractive force of the nucleus. The second electron affinity of oxygen involves the additon of an electron to a negatively charged species and is therefore endothermic. The first ionization of Cl^{2-}(g) brings about the formation of Cl^-(g) which has the electron configuration of argon and the formation of this stable ion is an exothermic process.

7 Answer **D**   *Comprehension of the Born–Haber cycle*
The only steps in the Born–Haber cycle, as written, which *must* be exothermic
are the first electron affinity of the halogen (s) and the lattice energy (t).

8 Answer **C**   *Comprehension of the Born–Haber cycle*
The only other steps which might be exothermic are the enthalpy changes of
formation (u) and solution (v). Whether these steps are exothermic depends on
the individual atoms involved. All the other steps are either definitely endo-'
thermic (p, q, r) or exothermic (s, t).

9 Answer **C**   *Comprehension of Hess's law diagrams*
In order to calculate $\Delta H_2^\ominus$, the standard enthalpy changes indicated by the
unlabelled arrow on the diagram need to be known. These are the standard
enthalpy changes of formation of $XO(g)$ and $Y_2O(g)$.

10 Answer **B**   *Knowledge of kinetic and energetic stabilities*
A substance is energetically stable with respect to a product if the standard free
energy change for the reaction, $\Delta G_{298}^\ominus$, is >0. It is kinetically stable if it shows no
tendency to react. Phosphorus is the only example given of a substance which is
unstable in both senses. It will react spontaneously with air forming an oxide and
the standard free energy change for the reaction is negative.

11 Answer **A**   *Knowledge of kinetic and energetic stabilities*
Of the four substances, gold shows no tendency to oxidize in air and the standard
free energy change of its oxidation is >0.

12 Answer **C**   *Comprehension of criteria for standard states*
Standard state implies the most stable form under the conditions specified.
Water is a liquid at $10\,°C$, 1 atm, as is bromine. Carbon dioxide is a gas under
these conditions. Hydrogen exists as diatomic gaseous molecules ($H_2(g)$) at
$10\,°C$, 1 atm.

13 Answer **B**   *Application of Hess's Law*
The answer is calculated using Hess's Law.

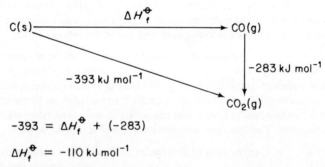

$$-393 = \Delta H_f^\ominus + (-283)$$

$$\Delta H_f^\ominus = -110 \text{ kJ mol}^{-1}$$

14 Answer **C**   *Knowledge of energy changes*
The process of sublimation (solid → gas) is endothermic because energy is
being taken in to raise the thermal energy of atoms/molecules. Similarly, the
ionization of calcium is endothermic because energy is taken in to remove the
electron. However, acceptance of an electron by a bromine atom to form a
stable bromide ion is exothermic.

15 Answer **B**   *Application of knowledge to experimental data*
25 cm^3 of 1 M NaOH contains $\frac{25}{100} = \frac{1}{40}$ mole of NaOH. The heat generated by its neutralization raises the temperature of 50 cm^3 of solution by 6.8 °C. 50 cm^3 of 0.5 M NaOH contains $\frac{50}{1000} \times 0.5 = \frac{1}{40}$ mole of NaOH. The expected tempereaure rise is only 3.4 °C because twice the volume of solution is being heated in this case.

16 Answer **B**   *Application of knowledge to experimental data*
The amount of heat liberated = $100 \times 4.18 \times 10$ J. This is liberated by complete combustion of $\frac{0.23}{46}$ moles. Therefore one mole will liberate

$$\frac{100 \times 4.18 \times 10 \times 46}{0.23} \text{ J.}$$

17 Answer **D**   *Knowledge of the Born–Haber cycle*
The enthalpy of formation for sodium chloride refers to it in its standard state (solid). Knowledge of its hydration energy is not required.

Knowledge of standard conditions necessary for the
18 Answer **B**   *measurement of heats of formation*
The heat of formation is defined as the enthalpy change when one mole of the compound in its standard state is prepared from its elements in their standard states, i.e. one mole of solid calcium fluoride is prepared from solid calcium and gaseous fluorine molecules under standard conditions of temperature and pressure.

Comprehension of the relationship between amount of
19 Answer **D**   *substance and enthalpy change*
The heat of combustion is the enthalpy change when one mole of the element/compound is completely burned in oxygen. The figure quoted is one quarter of the heat of combustion and this is evolved by one quarter of a mole of propane i.e. 5.6 dm^3 measured at s.t.p.

VI Equilibria

1 Answer B   *Comprehension of the units for an equilibrium constant*

The equilibrium constant for this reaction is:

$$K_c = \frac{[S(g)]^2[R(g)]}{[RS_2(g)]}.$$

The units for concentration are mol dm^{-3} and, therefore, the units for K_c are

$$\frac{(\text{mol dm}^{-3})^2(\text{mol dm}^{-3})}{(\text{mol dm}^{-3})} = (\text{mol dm}^{-3})^2 = \underline{\text{mol}^2 \text{ dm}^{-6}}.$$

2 Answer B   *Application of Le Chatelier's principle*

Decreasing the concentration of hydrogen will move the position of equilibrium to the left. Raising the temperature will move the position of equilibrium in the endothermic direction. Increasing the pressure will move the position of equilibrium in the direction having the smaller volume (NH$_3$—2 volumes; N$_2$ + 3 H$_2$— 4 volumes), increasing the concentration of ammonia. A positive catalyst will not alter the equilibrium position, or the concentration of ammonia, as it increases the rate of the forward reaction and the reverse reaction in the same proportion.

3 Answer B   *Application of Le Chatelier's principle*

An increase in pressure will only affect the equilibrium position of a reaction involving gases if there is a difference in volume between reactants and products. An increase in pressure moves the position of equilibrium in the direction having the smaller volume. The hydrogen/chlorine reaction involves no volume change (H$_2$ + Cl$_2$—2 volumes; 2 HCl—2 volumes). The volume of sulphur(VI) oxide (2 volumes) is smaller than that of sulphur dioxide and oxygen (3 volumes). In the other two examples there is an increase in volume if the underlined species is formed.

4 Answer C   *Application of knowledge of equilibrium constants*

In the reaction one mole of each reactant and one mole of each product is involved. The equilibrium constant is calculated from the concentrations at equilibrium. If x mole of C are formed in the reaction, x mole of D are also formed; x mole of A and x mole of B will have been used up. The concentration at equilibrium in mol dm^{-3} of each species is:

$$[A] = a - x \qquad [B] = b - x \qquad [C] = [D] = x$$

$$K_c = \frac{[C][D]}{[A][B]} = \frac{x^2}{(a-x)(b-x)}.$$

5 Answer **B**   *Application of knowledge of equilibrium constants*
   For this reaction:

$$K_p = \frac{(p_{NH_3})^2}{(p_{N_2})(p_{H_2})^3}$$

and the units are

$$\frac{(atm)^2}{(atm)(atm)^3} = atm^{-2}.$$

6 Answer **D**   *Knowledge of equilibrium constants*
   An equilibrium constant for a reaction remains constant unless the temperature
   is changed. In this reaction, increasing the pressure or using a catalyst does not
   change the constant. However, lowering the temperature will increase the
   concentration of Z and lower the concentration of Y.

$$K_c = \frac{[Z]^2}{[Y]}$$

                                           Application of the law of partial pressures to the
7 Answer **C**   *calculation of equilibrium constants*
   The partial pressures of the gases are in the same ratio as their mole fractions.
   The mole fractions are $N_2O_4 = \frac{1}{3}$; $NO_2 = \frac{2}{3}$.
   The partial pressure of $N_2O_4 = \frac{1}{3} \times 3 = 1$ atm.
   The partial pressure of $N_2 = \frac{2}{3} \times 3 = 2$ atm.

$$K_p = \frac{(p_{NO_2})^2}{(p_{N_2O_4})} = \frac{(2)^2}{1} = 4 \text{ atm.}$$

                                             Application of the law of partial pressures to
8 Answer **B**   *equilibrium situations*
   For this reaction $K_p = (p_{HCl})(p_{NH_3})$. Dissociation produces equal numbers of
   moles of ammonia and hydrogen chloride and they will have equal partial
   pressures. Therefore, $p_{HCl} = p_{NH_3} = \sqrt{K_p} = \sqrt{x}$ atm. The total pressure $=$
   $p_{HCl} + p_{NH_3} = \sqrt{x} + \sqrt{x} = \mathbf{2\sqrt{x}}$ **atm**.

9 Answer **B**   *Knowledge of solubility product expressions*
   The solubility product of the compound $A_a^{x+}B_b^{y+}(s)$ is equal to $[A^{x+}]^a[B^{y+}]^b$.
   The concentrations of each ion are raised to the powers a and b respectively,
   these numbers being equal to the number of moles of each individual ion in one
   mole of the compound. Hence, for $PbCl_2(s)$, the solubility product equals:

$$[Pb^{2+}(aq)][Cl^-(aq)]^2 \text{ mol}^3 \text{ dm}^{-9}.$$

10 Answer **D**   *Knowledge of equilibrium constants*
   In the reaction $aA(aq) + bB(aq) \rightleftharpoons cC(aq) + dD(aq)$

$$K_c = \frac{[C(aq)]^c[D(aq)]^d}{[A(aq)]^a[B(aq)]^b}$$

Product concentrations appear in the numerator and reactant concentrations in
the denominator. Each concentration is raised to the power equal to the number
of moles of substance appearing in the stoichiometric equation. In this reaction,

however, one of the products Ag(s) is in the solid state. Its concentration may be taken as constant because it is found to have no effect upon the concentrations of the aqueous species. Thus, its value is subsumed within the equilibrium constant itself.

11 Answer **D**   *Comprehension of the concentration of a gas*
The concentration of a gas can be expressed as the number of moles in one unit of volume. Hence, dividing n, the number of moles of gas, by V, the volume will generate the concentration in units of mol (volume)$^{-1}$. The units of R are usually expressed as $J\,K^{-1}\,mol^{-1}$.

12 Answer **C**   *Application of Le Chatelier's principle*
The addition of oxygen or nitrogen monoxide will move the position of equilibrium to the left. An increase in pressure will move the position of equilibrium in the direction of the smaller gas volume, i.e. $NO_2(g)$. Conversely a decrease in pressure will move the position of equilibrium in the direction of the larger gas volume, i.e. to the right.

13 Answer **C**   *Knowledge of heterogeneous equilibria*
In heterogeneous equilibria, all pure, condensed phases (pure solids and pure liquids) have a constant, non-zero concentration value. Consequently, in this example $K_c = [CO_2(g)]$.

14 Answer **A**   *Application of knowledge of partial pressures*
The mixture of gases is at 750 mmHg pressure. Oxygen comprises $\frac{1}{3}$ of the mixture by volume (its mole fraction is $\frac{1}{3}$) and, therefore, its resulting partial pressure is $\frac{1}{3} \times 750 = 250$ mmHg. The partial pressure is equal to the product of the mole fraction and the total pressure.

VII Kinetics

1 **Answer C**   *Application of knowledge to experimental kinetic data*
Comparing the data for reactions *a*, *b* and *c* it can be seen that the initial
concentrations of X and Y remain constant whilst that of W varies. Comparing
reactions *a* and *b*, as the initial concentration of W is doubled, the time taken for
the reaction to occur is approximately halved. Similarly, comparing reactions *a*
and *c*, as the initial concentration of W is increased threefold, the time taken for
the reaction to occur is reduced approximately threefold. The rate of reaction is
proportional to the concentration of W. Mathematically,

$$\text{Rate} \propto [W]^1$$

2 **Answer D**   *Application of knowledge to experimental kinetic data*
In reactions *a*, *d* and *e*, only the initial concentration of X is varying and the ratio
of the initial concentrations of X in these three reactions is $2:4:1$. However,
within experimental error, the times taken for the reactions to occur are equal. It
seems that changing the initial concentration of X and keeping the concentrations of W and Y constant has no effect upon the rate of reaction. Expressing
this mathematically,

$$\text{Rate} \propto [X]^0$$

3 **Answer B**   *Comprehension of the concept of order of reaction*
In reactions *a*, *f* and *g*, only the initial concentration of Y is varied. Comparing
reactions *a* and *f*, as the initial concentration of Y is doubled, the time taken for
the reaction to occur is approximately halved. Similarly, comparing reactions *a*
and *g*, as the initial concentration of Y is increased threefold, the time taken for
the reaction is reduced approximately threefold. The rate of reaction is proportional to [Y]. Mathematically,

$$\text{Rate} \propto [Y]^1$$

4 **Answer C**   *Comprehension of the concept of order of reaction*
The rate expression for the reaction is: rate $= k[W]^1[Y]^1$. The overall order of
the reaction is the sum of the powers to which the individual concentrations are
raised in the rate expression. In this case, $1 + 1$ and, therefore, the overall order
of the reaction is 2 (second order).

5 **Answer C**   *Application of knowledge to experimental kinetic data*
From questions one, two and three it follows that the rate of reaction is proportional to $[W]^1$, $[Y]^1$ and $[X]^0$. Putting constants of proportionality into these
expressions we have:

$$\text{rate} = k_1[W]^1; \qquad \text{rate} = k_2[Y]^1; \qquad \text{rate} = k_3[X]^0.$$

Combining these into an overall rate expression: rate $= k_1 k_2 k_3 [W]^1[Y]^1[X]^0$.
This simplifies to: rate $= k[W]^1[Y]^1$, where $k = k_1 k_2 k_3$ and $[X]^0 = 1$.

6 Answer **B** *Application of knowledge to experimental kinetic data*
The rate expression is: rate $= k[W]^1[Y]^1$.
The units for the left-hand side of the expression are mol $dm^{-3}s^{-1}$. The units for the right-hand side of the expression are (units of k) $mol^2\ dm^{-6}$. Therefore, from the equation it can be seen that the units of k are:

$$\frac{mol\ dm^{-3}\ s^{-1}}{mol^2\ dm^{-6}} = mol^{-1}\ dm^3\ s^{-1}.$$

7 Answer **D** *Knowledge of the terminology of kinetics*
The symbol k is given the name rate constant.

Application of knowledge of the equilibrium set up in
8 Answer **C** *the decomposition of hydrogen iodide*
This gas phase reaction involves no volume change and no ionic species are present. One of the products (iodine) is coloured, however. As the other participants in the equilibrium mixture are colourless, a colorimetric technique can be used to follow the course of the reaction.

9 Answer **D** *Comprehension of an activation energy diagram*
In the diagram, x is the activation energy and y is the enthalpy change. The products are energetically more stable than the reactants and, hence, the reaction is exothermic. Whether the reaction is exothermic or endothermic does not depend upon the relative magnitudes of x and y.

10 Answer **A** *Knowledge of the action of a catalyst*
A catalyst provides a different route for a reaction. The standard enthalpy change of the reaction remains unaltered. The catalyst effects a new, lower activation energy for the reaction by virtue of the new route.

Knowledge of the effect of a catalyst upon the
11 Answer **D** *standard enthalpy change of a reaction*
A catalyst has no effect upon the standard enthalpy change of a reaction. This is a constant at a particular temperature, being a function of the standard free energy change and the standard entropy change at that temperature:

$$\Delta G^{\ominus} = \Delta H^{\ominus} - T\Delta S^{\ominus}.$$

The route that a reaction proceeds along does not affect this.

Comprehension of the unimolecular
12 Answer **B** *hydrolysis of bromoalkanes*
Reaction (i) is an ionization and, hence, is the slower of the two steps. For this reason, it is known as the rate determining step of the reaction. In other words, the rate of the whole reaction is determined by this, the slower, reaction. Reaction (ii) involves the combination of an ion R^+ with a polar molecule H_2O and is relatively fast. Increasing the hydroxide ion concentration in this reaction would increase the rate as now two ionic species would be involved, and ionic reactions are generally very fast. Experiments show that the rate expression for this kind of hydrolysis is: rate $= k[RBr]$. It is important to remember that this information cannot be deduced from the equations given in the question. It can only be determined experimentally.

13 Answer **A**   *Comprehension of kinetic data in a graphical form*

14 Answer **C**

15 Answer **B**

In a series of reactions which are first order with respect to R, the initial rate of reaction is directly proportional to the concentration of R, i.e. rate = k[R]. This is a linear relationship (graph A). The gradient of the graph is the rate constant k. When the order of a reaction is zero with respect to a reactant, increasing the concentration of the reactant has no effect upon the rate (graph C). In a first order reaction, the concentration of reactant falls exponentially as the time elapsed increases (graph B). The rate of reaction can be determined at any time by taking the gradient of the curve at that time. These gradients plotted against concentration will give a straight line graph (gradient = rate constant) only for a first order reaction.

16 Answer **C**   *Knowledge of the 'terms order' and 'overall molecularity'*
The overall molecularity of a reaction is the sum of the stoichiometric coefficients of the starting materials in a reaction. In the reaction: $2 X + 3 Y_2 \rightarrow$ products, the overall molecularity is $2 + 3 = 5$. The rate expression and the order of reaction with respect to any reactant can only be determined by experiment and cannot be predicted from the equation.

17 Answer **B**   *Knowledge of the mode of operation of catalysts*
By definition, a catalyst is chemically unchanged at the end of a reaction; it *may*, however, be physically changed. Most catalysts are believed to act by forming an intermediate which possesses lower energy than the intermediate formed in the uncatalyzed reaction. The reaction route is altered.

18 Answer **C**   *Knowledge of a photochemical reaction*
The rate of some reactions may be altered by using ultra-violet radiation (bright sunlight is sometimes suitable as a source). In this example, the first step is the homolytic fission of a chlorine–chlorine bond to form two chlorine free radicals. Subsequently, these react with hydrogen molecules to form hydrogen chloride.

19 Answer **C**   *Knowledge of the effects of temperature upon rates of reaction*
The collision theory of reaction kinetics states that there must be an effective collision between molecules before a reaction will take place. That is to say that the colliding molecules must possess a minimum energy and must be correctly aligned with respect to each other. This minimum energy is the activation energy. If the rate is increased fourfold, this is because there are four times as many molecules as before possessing the minimum energy.

20 Answer **B**   *Application of data in a rate expression*
If the concentration of reactant A is doubled, the rate will increase by a factor of $2^2 = 4$ because the reaction is second order with respect to A. If the concentration of B is increased fourfold, the rate will change by a factor of $4^{1/2} = 2$. The reaction is of order $\frac{1}{2}$ with respect to B. The overall effect upon the rate, therefore, is an increase by a factor of $4 \times 2 = 8$.

21 Answer **B**   *Application of graphical information on reaction rates*
The straight line graph shows that the reaction rate is directly proportional to the
initial concentration of $N_2O_5(g)$. The reaction is of first order with respect to
$N_2O_5(g)$: rate $= k[N_2O_5(g)]$. The gradient is equal to k, the rate constant, which
is a constant only when the temperature is constant. The units of k are:

$$\frac{\text{mol dm}^{-3}\,\text{s}^{-1}}{\text{mol dm}^{-3}} = \text{s}^{-1}.$$

VIII Colligative Properties

1 Answer D   *Comprehension of the term 'colligative property'*
A colligative property is one which depends solely on the *number* of particles in
a given volume of solvent. It can be used to find the relative molecular mass of a
solute. pH is not a colligative property as it is dependent on the number of
hydrogen ions present, rather than the total number of ions present.

2 Answer C   *Knowledge of the cause of the lowering of vapour pressure*
The vapour pressure of a solvent is reduced when a non-volatile solute is added,
because the rate at which solvent molecules leave the surface is reduced by the
presence of solute molecules. The rate at which solvent molecules return to the
surface from the vapour is not decreased. It is the reduction of vapour pressure
which causes the boiling point to be raised.

3 Answer D   *Knowledge of Raoult's law*
Raoult's law states that the relative lowering of vapour pressure is equal to the
mole fraction of the solute, i.e.

$$\frac{\text{lowering of vapour pressure}}{\text{vapour pressure of solvent}} = \frac{\text{number of moles of solute}}{\text{total number of moles in solution}}$$

Therefore

$$\frac{p_0 - p_1}{p_0} = \frac{n}{N+n}.$$

Where

$$p_0 > p_1.$$

4 Answer A   *Application of Raoult's law*
In a *very* dilute solution, when the number of solute molecules present is very
small $N + n \approx N$, and

$$\text{the relative lowering of vapour pressure} = \frac{\text{number of moles of solute}}{\text{number of moles of solvent}} = \frac{n}{N}$$

$$0.05 = \frac{n}{4} \qquad n = 0.2.$$

5 Answer C   *Comprehension of the term mole fraction*

$$\text{mole fraction} = \frac{\text{number of moles of solute}}{\text{total number of moles present in solution}}$$

$$= \frac{\frac{2}{4000}}{\frac{37}{148} + \frac{2}{4000}}.$$

6 Answer **B**
Equal numbers of moles of different non-volatile solutes cause the same elevation of boiling point in equal quantities of the same solvent. There are $\frac{3}{60}$ (0.05) moles of solute X dissolved in 100 cm^3 of solution. The equivalent concentration is 0.1 moles in 200 cm^3. Hence 6 g of solute Y is equivalent to 0.1 moles. The relative molecular mass of Y is 60.

7 Answer **D**   *Comprehension of the effect of dissociation on colligative properties*
The amount by which the boiling point of a solution is raised is dependent solely on the number of solute particles present, unless dissociation occurs. Glucose is undissociated and copper(II) oxide remains undissolved. One mole of sodium chloride dissociates giving two moles of ions, but one mole of magnesium chloride dissociates to yield three moles of ions. This produces the largest concentration of particles and therefore the largest colligative effect.

8 Answer **B**
Vapour pressure is dependent upon the temperature of the solution. The relative lowering of vapour pressure (to that of pure solvent) is dependent on the number of particles from the solute which are present. This is influenced by the degree to which the solute dissociates. It is not related to the melting point of the solute.

9 Answer **D**   *Comprehension of the term 'ebullioscopic constant'*
The ebullioscopic constant is the amount by which the boiling point is raised if one mole of undissociated, non-volatile solute is dissolved in 1 kg of solvent. Hence one mole of glucose raises the boiling point of 1 kg of water by 0.52 °C and 0.5 mole in 1 kg of water will cause half the elevation. However, one mole of sodium chloride is dissociated into two moles of ions, resulting in an elevation of 1.04 °C if 1 kg of water is used.

10 Answer **C**
3 g in 100 g cause an elevation of 0.54 °C, therefore 3 g in 1000 g will cause an elevation of 0.054 °C. 1 mole in 1000 g causes an elevation of 2.7 °C and so

$$\text{No. of moles} = \frac{0.054}{2.7} = \frac{2}{100}.$$

3 g is 0.02 mole and so the relative molecular mass of X $= \frac{3}{0.02} = 150$.

11 Answer **D**   *Application of elevation of boiling point data*
13 g of X is $\frac{13}{65} = 0.2$ mole.
0.2 mole in 200 cm^3 causes an elevation of 0.37 °C,
therefore 0.4 mole in 400 cm^3, also causes an elevation of 0.37 °C.
The mass of X required $= 124 \times 0.4$ g $= 49.6$ g

12 Answer **C**   *Comprehension of experimental data*
The vapour pressure of a solution is less than that of the pure solvent at the same temperature, and so curve **B** represents the solution and curve **A** the pure solvent. The boiling point is raised from T_A to T_B by the addition of solute

because the boiling point is the temperature at which the vapour pressure is equal to the external pressure (1 atm in this question).

$$\text{The relative lowering of vapour pressure} = \frac{1 - p_B}{p_B}.$$

13 Answer C
Comprehension of the effect of association and dissociation on colligative properties
The true value of the relative molecular mass is obtained if the solute is neither associated nor dissociated. Ethanoic acid exists as a dimer in a non-polar solvent due to intermolecular hydrogen bonding. Both potassium chloride and ethanoic acid dissociate into ions when dissolved in water, the former is fully ionized but the latter is only partly ionized.

14 Answer B
Knowledge of the process of osmosis
Osmosis is the transfer of solvent molecules across a semi-permeable membrane to a region of higher solute concentration. The process tends to equalize concentrations between solutions. A semi-permeable membrane does not allow the passage of solute molecules.

15 Answer C
Comprehension of the effect of concentration on osmotic pressure
Any solution of one mole of undissociated solute in 22.4 dm^3 of solution is defined as having an osmotic pressure of 1 atm at 0 °C. Solutions **A**, **C** and **D** each have a concentration equivalent to one mole in 22.4 dm^3. Each mole of potassium iodide is dissociated into two moles of ions and will result in an osmotic pressure of 2 atm.

16 Answer C
Comprehension of the effect of dissociation on osmotic pressure
Solutions of equal concentrations will have equal osmotic pressures unless the solute is dissociated. Osmotic pressure, a colligative property, is dependent on number of particles present in the solution. Sulphuric acid in dilute solution dissociates into three ions per molecule and therefore has the greatest osmotic pressure.

17 Answer D
Application of a colligative property to find the relative molecular mass of a solute
24 g of X in 11.2 dm^3 produces an osmotic pressure of 0.5 atm. Therefore 24 g of X in 22.4 dm^3 will produce an osmotic pressure of 0.25 atm. The relative molecular mass of X is 96 because 96 g dissolved in 22.4 dm^3 would produce an osmotic pressure of 1 atm.

18 Answer C
Comprehension of conditions necessary for ideal behaviour in liquid mixtures
For a mixture of liquids to be ideal the components should be as closely related chemically, as possible. This minimizes the bonding changes which occur between the components on mixing. If hydrogen bonding or dipole–dipole interaction is present between the components it will cause lowering of vapour pressure. The substances in response **C** are both six-membered hydrocarbons and are chemically most similar.

19 Answer **D**

Application of a colligative property to find the degree of dissociation of a solute

If the degree of dissociation is α, one mole of HA will yield α moles of H^+, α moles of A^- and $1-\alpha$ moles of HA when equilibrium is reached. The total number of moles is $1+\alpha$. The expected depression is $0.3\,°C$ but the actual depression $(0.4\,°C)$ is higher due to dissociation.

$$\frac{1+\alpha}{1} = \frac{0.4}{0.3} \quad \text{and} \quad \alpha \text{ (the degree of dissociation)} = \frac{0.4}{0.3} - 1.$$

IX Acid/Base Reactions

1 Answer **C**   *Comprehension of the meaning of pH*
pH can be defined as the negative logarithm, to the base 10, of the hydrogen ion concentration in the solution/liquid, with the concentration measured in $mol\,dm^{-3}$. This is usually written: $pH = -\log_{10}[H^+]$, but may also be written: $pH = \log_{10}\dfrac{1}{[H^+]}$. Similarly, $pOH = -\log_{10}[OH^-]$ and at 25 °C, $pH + pOH = 14$. It follows that $pH = 14 - pOH = 14 + \log_{10}[OH^-]$.

2 Answer **C**   *Application of knowledge about pH*
Change in $pH = -\log_{10}$ (change in hydrogen ion concentration). It follows that a change in pH of three units is caused by a change in $[H^+]$ by 10^3. A rise in pH shows increasing alkalinity and a reduction in $[H^+]$.

3 Answer **C**   *Application of knowledge about pH*
Employing the same equation as in answer two, a change in $[H^+]$ by a factor of 10^2 causes a change in pH of 2 units. Increasing the hydrogen ion concentration leads to a lowering of pH.

4 Answer **B**   *Comprehension of the ionic product of water*
The ionic product of water, $K_w = [H^+][OH^-]\,mol^2\,dm^{-6}$. The ions arise from the dissociation of water: $H_2O(l) \rightleftharpoons H^+(aq) + OH^-(aq)$. It follows that in pure water, $[H^+] = [OH^-]$ and that $K_w = [H^+]^2$. At 70 °C, therefore, $[H^+] = \sqrt{K_w} = 1 \times 10^{-6}\,mol\,dm^{-3}$, and $pH = 6$. The pH of pure water is 7 only at 25 °C.

5 Answer **D**   *Comprehension of the ionic product of water*
The ionic product of water, $K_w = [H^+][OH^-]\,mol^2\,dm^{-6}$. The ions arise from the dissociation of water. As the temperature increases, the value of K_w increases showing that the position of equilibrium $H_2O(l) \rightleftharpoons H^+(aq) + OH^-(aq)$ moves to the right as the temperature increases. A reaction whose equilibrium constant increases with increasing temperature is an endothermic one. At 30 °C, $[H^+] = \sqrt{K_w} = \sqrt{4 \times 10^{-14}} = 2 \times 10^{-7}\,mol\,dm^{-3}$. As the hydrogen ion concentration at this temperature is greater than 10^{-7}, the pH of the solution will be less than 7.

6 Answer **D**   *Comprehension of the relationship between concentration and pH*
A monobasic acid, when fully ionized, will give one mole of hydrogen ions and one mole of anions per mole of acid. Thus, the hydrogen ion concentration is equal to the original acid concentration. In this case $pH = -\log_{10}(c) = -1$ (where c is the concentration of the acid/$mol\,dm^{-3}$). However, at this concentration, all known strong acids have a pH less than expected from simple calculations such as this, because the effective concentration of hydrogen ions is less than would be expected from concentration considerations alone.

7 Answer **D**   *Application of knowledge of weak acids*
If the monobasic acid is 10% dissociated, one mole of the acid will generate 0.1 mole of hydrogen ions. The concentration of hydrogen ions, therefore, is equal

to $0.01 \times 0.1 = 0.001$. Therefore, $[H^+] = 10^{-3}$ mol dm^{-3}. The pH value of the solution will be $-\log_{10}(10^{-3}) = 3$.

8 Answer A   *Comprehension of the definition of a Lowry–Brönsted acid*
A substance is classed as a Lowry–Brönsted acid if it acts as a proton donor. In this example, sulphuric acid is acting as an acid by the formation of the hydrogensulphate ion (HSO_4^-). Nitric acid is acting as a base by accepting protons to form H_3O^+ and NO_2^+.

9 Answer B   *Application of the relationship between pH and concentration*
In this reaction, 1 cm^3 of the 0.1 M HNO_3 remains not neutralized. The total volume of the solution is 99 cm^3 and so the concentration of hydrogen ions in the solution is equal to $\frac{1}{99} \times 0.1 \approx 0.001$ mol dm^{-3}. The approximate pH of the solution, therefore, is $-\log_{10}(10^{-3}) = 3$. Although 98% of the acid has been neutralized, the pH change has only been approximately 2 units. Had 49.9 cm^3 of 0.1 M alkali been used, the pH change would have been approximately 3 units. It is because of considerations such as these that there is a very rapid change in pH near the neutral point of a titration between a strong acid and a strong base. Consequently, the titration curves for such systems have their characteristic shape and indicators give sharp end-points.

10 Answer D   *Knowledge of salt hydrolysis*
The salt of a strong acid and strong base will form a neutral solution in water. The salt of a weak acid and strong base e.g. sodium carbonate, is alkaline in solution. The following equilibrium is set up:

$$CO_3^{2-}(aq) + 2\,H_2O(l) \rightleftharpoons H_2CO_3(aq) + 2\,OH^-(aq)$$

The removal of hydrogen ions from the water results in an excess of hydroxide ions and the solution is alkaline. A solution of ammonia is alkaline. A solution of aluminium chloride in water contains the hydrated aluminium(III) ion $[Al(H_2O)_6]^{3+}$, and the following equilibrium is set up:

$$H_2O + [Al(H_2O)_6]^{3+} \rightleftharpoons [Al(H_2O)_5(OH)]^{2+} + H_3O^+$$

The solution is acidic owing to the formation of H_3O^+ ions.

11 Answer C   *Application of the relationship between K_a and pK_a*
By definition, p$K_a = -\log_{10}K_a$. It follows that the pK_a value for propanoic acid is $-\log_{10}(10^{-5}) = 5$, and the pK_a value for trichloroethanoic acid is $-\log_{10}(10^{-1}) = 1$. The ratio required in the question, therefore, is $\frac{5}{1}$.

12 Answer B   *Comprehension of the relationship between the basicity of an acid and its pH in solution*
pH $= -\log_{10}[H^+]$ and, therefore, if the pH is 2, the hydrogen ion concentration will be equal to 10^{-2} mol dm^{-3}. This is not influenced by the fact that the acid is dibasic (H_2SO_4); however, assuming that the acid is fully ionized, the concentration of the acid must be half that of the hydrogen ions coming from the acid. The concentration of the acid is 0.5×10^{-2} mol dm^{-3}.

13 Answer C   *Comprehension of the expressions for K_a and pK_a*
For this reaction, by definition,

$$K_a = \frac{[Cl_3CCOO^-][H_3O^+]}{[Cl_3CCOOH]}.$$

The concentration of water is approximately constant in dilute solution and its value is subsumed within K_a.

$$pK_a = -\log_{10}K_a$$

and, therefore, the expression for pK_a for this reaction is:

$$-\log_{10}\left(\frac{[Cl_3CCOO^-][H_3O^+]}{[Cl_3CCOOH]}\right)$$

which may also be written as:

$$\log_{10}\left(\frac{[Cl_3CCOOH]}{[Cl_3CCOO^-][H_3O^+]}\right).$$

14 Answer A *Comprehension of redox and acid/base reactions*
In reaction (a), a proton has been donated from the water to the ammonia forming the ammonium ion and the hydroxide ion. In this reaction, ammonia is acting as a base (proton acceptor). In reaction (b), ammonia both accepts a proton and donates a proton forming NH_4^+ and NH_2^- respectively. In this case it is acting both as a base and an acid (proton donor).

Application of knowledge about the ionic
15 Answer C *product of water and pH*
In 0.001 M sodium hydroxide solution $[OH^-] = 10^{-3}$ mol dm^{-3}. At 25 °C,

$$K_w = [OH^-][H^+] = 10^{-14} \text{ mol}^2 \text{ dm}^{-6}$$

and it follows that:

$$[H^+] = \frac{10^{-14}}{[OH^-]} = \frac{10^{-14}}{10^{-3}} = 10^{-11} \text{ mol dm}^{-3}.$$

$$\underline{pH = 11.}$$

Knowledge of the dependence of the pH of a solution
16 Answer B *upon the degree of ionization of the solute*
The assumption that has to be made is that the solute is completely ionized in the solution. If this were not the case, the hydroxide ion concentration would be smaller than expected and, consequently, the hydrogen ion concentration would be greater than expected and the pH would be lower. The rest of the information given in the question is irrelevant.

Knowledge of conditions necessary for buffering
17 Answer B *to occur within a solution*
In order for a mixture of chemicals to have a buffering action within a solution, they must be either a mixture of a weak acid and a salt that it forms with a strong base (examples **A** and **D**) or the converse, a weak base and the salt that it forms with a strong acid (example **C**). Example **B** is a mixture of a strong acid and the salt that it forms with a strong base.

18 Answer C *Knowledge of the properties of hydrochloric acid*
At high concentrations of hydrogen chloride gas in water, there is incomplete ionization; the larger the concentration becomes, the smaller is the degree of

ionization. In theory, acidic solutions whose hydrogen ion concentrations are in excess of 1 mol dm^{-3} should have a pH $<$ 0. For example, a 2 M solution of a fully ionized monobasic acid should have, in theory, a pH of $(-\log_{10}(2)) = -0.3$. Although negative pH values are found at high hydrogen ion concentrations, incomplete ionization ensures that pH values are always more positive than simple calculation would have us believe.

19 Answer **C**   *Application of knowledge of dissociation constants*
The overall dissociation constant for this reaction is:

$$K_a = \frac{[A^{2-}][H^+]^2}{[H_2A]}.$$

Observation tells us that this quotient is generated by multiplying K_i by K_{ii}:

$$K_i \times K_{ii} = \frac{[HA^-][H^+] \times [A^{2-}][H^+]}{[H_2A][HA^-]} = \frac{[A^{2-}][H^+]^2}{[H_2A]} = K_a.$$

20 Answer **C**   *Application of knowledge about weak acids*
The two substances have different K_a values and are ionized to different extents. This results in different hydrogen ion concentrations and different values of pH in equimolar solutions. Phenol has the smaller K_a value and is therefore a weaker acid than ethanoic acid. Phenol has a larger pK_a value than ethanoic acid because of the inverse logarithmic relationship between K_a and pK_a.

X Redox Reactions

1 Answer **D** *Comprehension of the concept of oxidation number*
In a compound, ion or other atomic species, the sum of the oxidation numbers of the constituent atoms must equal the charge on the species. The sum in uncharged compounds is zero, in ions with a charge of one positive it is +1, etc. In compounds other than peroxides, oxygen has an oxidation number of −2 and hydrogen, in compounds other than metallic hydrides, has an oxidation number of +1. From this, it can clearly be seen that chromium has an oxidation number of +6 in species **A**, **B**, and **C**:

A $(+6) + 4(-2) = -2$
B $2(+6) + 7(-2) = -2$
C $(+6) + 3(-2) = 0$

In species **D**, however, the oxidation number of chromium is +3:

D $(+3) + 3(-2 + 1) = 0$

2 Answer **A** *Comprehension of the relationship of oxidation number to electron transfer*
For every electron transferred per atom, there is a change in oxidation number of one. For example, in **A**, manganese is reduced from oxidation state +7 (manganate(VII)) to oxidation state +2 (manganese(II)) and this is equivalent to a five electron reduction. The changes occurring in the other reactions are:

B 2 Cr(VI) to 2 Cr(III); a six electron reduction (2×3)
C Mn(VI) to Mn(IV); a two electron reduction
D Cr(VI) to Cr(III); a three electron reduction

3 Answer **C** *Comprehension of the relationship of oxidation number to redox reactions*
If an atom has an increase in oxidation number during a reaction, it is being oxidized. If a decrease in oxidation number occurs, the atom is being reduced. If both an increase and a decrease occurs, the change is classed as disproportionation. In reaction **C**, no change of oxidation number occurs for chromium, oxygen or hydrogen; the chromium remains in oxidation state +6 throughout the reaction.

Reaction	Oxidation	Reduction
A	Fe(II) to Fe(III)	Mn(VII) to Mn(II)
B	I(−I) to I(0)	Cl(0) to Cl(−I)
D	Cu(0) to Cu(II)	S(VI) to S(IV)

4 Answer **D** *Knowledge of disproportionation*
Disproportionation is the simultaneous oxidation and reduction of the same atomic species. Inspecting the reactions:

Reaction	Reactant		Products
A	2Ti(III)	to	Ti(II) + Ti(IV)
B	3Mn(VI)	to	2Mn(VII) + Mn(IV)
C	2Cl(O)	to	Cl(−I) + Cl(I)
D	2S(VI)	to	S(VI) + S(IV)

In example **D**, the oxidation number of sulphur does not increase during the reaction, although it does fall. Only reduction of sulphur is occurring here and it is the copper which is being oxidized.

5 Answer **C**   *Application of knowledge to the balancing of redox equations*
The equation for the oxidation of ethanedioate (oxalate) ions is:

$$C_2O_4{}^{2-}(aq) \rightarrow 2\,CO_2(g) + \underline{2\,e^-}.$$

The equation for the reduction of manganate(VII) ions in acidic solution is:

$$8\,H^+(aq) + MnO_4{}^-(aq) + \underline{5\,e^-} \rightarrow Mn^{2+}(aq) + 4\,H_2O(l).$$

If the equations are put together and balanced to allow the transfer 10 electrons, we have:

$$16\,H^+(aq) + 5\,C_2O_4{}^{2-}(aq) + 2\,MnO_4{}^-(aq) \rightarrow$$

$$2\,Mn^{2+}(aq) + 8\,H_2O(l) + 10\,CO_2(g).$$

Five moles of ethanedioate ions will react with two moles of manganate(VII).

6 Answer **D**   *Knowledge of the oxidizing reactions of concentrated sulphuric acid*
The reaction of concentrated sulphuric acid with sodium chloride results in the formation of hydrogen chloride gas which is stable to oxidation by the concentrated acid. There is no oxidation or reduction occurring here. The neutralization reaction with sodium hydroxide and the reaction with sodium nitrate ($NaNO_3$) to form nitric acid are not redox reactions either. When sodium bromide reacts with the concentrated acid, hydrogen bromide is formed which is subsequently oxidized to bromine, reducing the sulphuric acid to sulphur dioxide.

7 Answer **D**   *Knowledge of the redox reactions of hydrogen peroxide*
In acidic solution, manganate(VII) ions act as powerful oxidizing agents being reduced in the process to manganese(II) ions. The reaction of hydrogen peroxide with manganate(VII) ions in acidic solution results in the formation of oxygen by the oxidation of the peroxide. Manganate(VII) is a stronger oxidizing agent than peroxide in acidic solution. The relevant redox potentials are:

$$[2\,H^+(aq) + O_2(g)],\ H_2O_2(aq)|Pt; \qquad E^\ominus = +0.68\ V$$

$$[MnO_4{}^-(aq) + 8\,H^+(aq)],\ [Mn^{2+}(aq) + 4\,H_2O(l)]|Pt; \qquad E^\ominus = +1.51\ V$$

8 Answer **A**   *Comprehension of changes in oxidation number*
Calcium hydride is ionic, containing Ca^{2+} and H^- ions. In the formation of calcium hydride from its elements, the metal is oxidized and the hydrogen is reduced. With fluorine and oxygen, the hydrogen is oxidized in both cases and with ethyne, C_2H_2, we have reduction to ethene, C_2H_4, by the simple addition of hydrogen.

9 Answer **D**   *Knowledge of the reactions of the compounds of iron*
The displacement of iron by magnesium from iron(II) sulphate solution involves
the change: Fe(II) to Fe(0). Iron(III) chloride is reduced to iron(II) chloride by
hydrogen sulphide and hydrogen gas reduces Fe_3O_4 (iron(II) diiron(III) oxide)
to iron metal. In the reaction of sodium hydroxide with iron(III) nitrate,
however, iron(III) hydroxide is precipitated and there is no change in the
oxidation number of the iron.

10 Answer **C**   *Knowledge of maximum oxidation numbers*
The maximum oxidation number of an element is usually equal to the number of
electrons which can be removed from the outer level of the atom. For elements
in Groups I to VII it corresponds with the group number of the element.
Examples of maximum oxidation numbers are: sulphur +6; magnesium +2;
lead +4 and sodium+1. Oxygen and fluorine are the best known exceptions to
this general rule. In the examples given, ClO_2 is the only one which contains an
element which is not in its maximum oxidation state. The oxide of chlorine
Cl_2O_7 exists.

11 Answer **B**   *Comprehension of changes in oxidation state*
In the reaction copper(II) is reduced to copper(I). Some of the iodine remains
unchanged in the (−I) oxidation state, whilst some is oxidized to the element:

$$2\,I^- \rightarrow I_2 + 2\,e^-.$$

12 Answer **B**   *Comprehension and application of $E^\ominus$ values*
The $E^\ominus$ value of a standard half-cell shows its oxidizing and reducing power. For
example, the more negative the $E^\ominus$ value is, the better reducing agent the metal
is and, generally, the more reactive it is and the better oxidizing agent the metal
ion is. The element with the largest negative $E^\ominus$ value has the greatest tendency
to exist in solution as the positive ion (has the greatest tendency to be oxidized).
It is, therefore, the strongest reducing agent. It follows that potassium is the
strongest reducing agent of the four metals and iron *can* displace silver from
solution. Zinc can be used as a protective coating for iron or as a sacrificial anode
because it will be oxidized in preference to iron. Silver metal is the weakest
reducing agent listed. Note that tin can also be used as a protective coating for
iron (tin plate) but in this case, the tin will not oxidize preferentially to the iron
because of the positions of the appropriate redox potentials. When the tin
surface is broken, the iron will rust.

13 Answer **C**   *Knowledge of changes occurring in an electrochemical cell*
In the Daniell cell, copper is deposited at the copper electrode by the reduction
of copper(II) ions:

$$Cu^{2+}(aq) + 2\,e^- \rightarrow Cu(s).$$

Electrons are provided by zinc dissolving at the surface of the zinc electrode:

$$Zn(s) \rightarrow Zn^{2+}(aq) + 2\,e^-.$$

The electrons flow round the external circuit to the copper electrode. The zinc
rod becomes surrounded by Zn^{2+} ions and sulphate ions migrate towards it.

14 Answer **D** *Application of $E^{\ominus}$ values*

By convention, the e.m.f. of a cell is given by the equation:

$$E^{\ominus}_{cell} = E^{\ominus}_{(right\text{-}hand\ half\text{-}cell)} - E^{\ominus}_{(left\text{-}hand\ half\text{-}cell)}$$

In this case:

$$E^{\ominus} = 0.34 - (-0.76) = +1.1 \ V.$$

Clearly, the amount of energy derivable from a cell is the same no matter which way round the cell diagram is written, but the potential of the cell may be positive or negative. For example, had the diagram of the Daniell cell been written in the reverse order with the copper half-cell on the left-hand side instead of on the right, the e.m.f. would have been -1.1 V. The sign of the e.m.f. of a cell is equal to the polarity of the right-hand electrode.

15 Answer **C** *Comprehension of half-cells and redox potentials*

The diagram shows two oxidizing systems:

(i) $[MnO_4^-(aq) + 8\ H^+(aq)]$ (ii) $[2\ H^+(aq) + S(s)]$

and two reducing systems:

(iii) $[Mn^{2+}(aq) + 4\ H_2O(l)]$ (iv) $H_2S(aq)$.

Under acidic conditions, the manganese system is the more powerful oxidant (i) because it has the more positive $E^{\ominus}$ value and the hydrogen sulphide system (iv) is the more powerful reductant because it has the most negative $E^{\ominus}$ value. It follows that these two systems will react together and the position of equilibrium will lie to the right. The manganese system is reduced and the hydrogen sulphide is oxidized. Because hydrogen ions occur in the half-cell expressions, their concentrations will affect the half-cell potentials and, in this case, the whole cell potential.

16 Answer **D** *Application of knowledge of cell diagrams*

17 Answer **A**

18 Answer **B**

In order to calculate the e.m.f. of a cell from a knowledge of the half-cell potentials and the cell diagrams, it is important to realize that the half-cells as shown are presented as the right-hand component of a complete cell which has the standard hydrogen electrode (unwritten) as the left hand part. The e.m.f. of a complete cell is found by subtracting the $E^{\ominus}$ value of the left-hand half-cell from that of the right-hand half-cell:

16 $E^{\ominus} = (0.4) - (-1.2) = +1.6 \ V$

17 $E^{\ominus} = (-0.6) - (-0.8) = +0.2 \ V$

18 $E^{\ominus} = (0.4) - (-0.2) = +0.6 \ V.$

In each case $E^{\ominus} > 0$ signifying that the right-hand half-cell has positive polarity with respect to the left hand one.

19 Answer **D**

The cell shown is made up of two standard half-cells. The reactions occurring in them are:

(a) $\quad\quad\quad Zn(s) \rightleftharpoons Zn^{2+}(aq) + 2\,e^-; \quad\quad E^{\ominus} = -0.76 \text{ V}.$

(b) $\quad\quad\quad Cu^{2+}(aq) + 2\,e^- \rightleftharpoons Cu(s); \quad\quad E^{\ominus} = +0.34 \text{ V}.$

Altering the concentration of zinc ions in the zinc half-cell will have no effect upon the e.m.f. of the copper half-cell as they function independently, although it will affect the e.m.f. of the cell as a whole. By Le Chatelier's principle, lowering the concentration of zinc ions will result in the position of equilibrium in reaction (a) moving from left to right causing the availability of electrons from the cell to increase and, hence, cause the half-cell e.m.f. value to become more negative. Therefore, the $E^{\ominus}$ value of the cell as a whole will now be represented by: $E^{\ominus} = 0.34 - (-x)$ volts, where $-x < -0.76$. It follows that $E^{\ominus}_{cell}$ will now be more positive than 1.1 V. Also by Le Chatelier's principle, raising the copper(II) ion concentration will make that half-cell e.m.f. value more positive and this will make the $E^{\ominus}_{cell}$ value more positive also. Thus, lowering the zinc ion concentration and lowering the copper ion concentration will both, independently, make the $E^{\ominus}_{cell}$ value more positive.

20 Answer **C** $\quad\quad\quad\quad\quad\quad$ *Knowledge of standard redox potentials*

With the e.m.f. of the standard hydrogen electrode defined as zero volts, any standard half-cell having a greater tendency to release electrons than the standard hydrogen half-cell will have a negative potential with respect to the hydrogen system. Those half-cells with a lesser tendency will possess a positive potential with respect to the hydrogen system. In this question, only the zinc system has a negative e.m.f. (-0.76 V).

21 Answer **C** *Comprehension of a graphical representation of the Nernst equation*

Plotting $E^{\ominus}$ values against $\log_{10}K_c$ at constant temperature, will result in a straight line graph with gradient $2.3RT/zF$. For a reaction at 298 K with a two electron transfer, the gradient has a value of 0.03 V. The gradient is always positive.

XI Groups I, II and III

1 Answer **A**   *Knowledge of reactions of Group II elements*
The reactivity within a group of metals increases down the group. Calcium, strontium and barium will react with both water and steam. Beryllium does not react with water. Formation of Be^{2+} is energetically difficult owing to the high ionisation energies involved.

2 Answer **D**   *Knowledge of reactions of ionic hydrides*
The hydrides of Group I and II metals are mainly ionic, except for those of beryllium and magnesium. Ionic hydrides react with water to give hydrogen, e.g.
$$NaH(s) + H_2O(l) \rightarrow NaOH(aq) + H_2(g).$$

3 Answer **A**   *Knowledge of solubilities of Group I halides*
Lithium fluoride is the only Group I fluoride which is insoluble. This is explained by the very large value of the lattice energy of lithium fluoride, which more than compensates for the large heats of hydration of lithium and fluoride ions.

4 Answer **B**   *Knowledge of Group I oxides*
The common formula for a Group I oxide is M_2O. However, peroxides are known, having a formula M_2O_2 (not known for lithium). Potassium, rubidium and caesium also form superoxides (MO_2). The structure of the superoxide seems to be ionic, involving the O_2^- ion. The peroxide structure is also ionic and involves the O_2^{2-} ion.

5 Answer **D**   *Knowledge of the properties of Group I hydroxides*
The base strength increases down the hydroxides of Group I. The attraction between large cations and the hydroxide ion is less than that between small cations and the hydroxide ion. The size of the cation increases down Group I.

6 Answer **C**   *Knowledge of the thermal stability of Group I and II compounds*
Hydrogencarbonates are generally unstable when heated. Carbonates, formed by metal ions which have large radius to charge ratios are stable to heating. Those formed by metal ions with low radius : charge ratios, e.g. Li^+, are unstable because the small metal ion polarizes the carbonate ion.

7 Answer **B**   *Knowledge of the decomposition of Group I and II nitrates*
Calcium, magnesium and lithium nitrates decompose on heating to give the metal oxide, oxygen and nitrogen dioxide. Group I nitrates, other than that of lithium, produce a solid nitrite (e.g. $NaNO_2$) and oxygen gas. This is another anomalous property of lithium which arises from the large polarising power of the lithium cation.

Application of knowledge concerning the hydrolysis
8 Answer **C**   *of $[Al(H_2O)_6]^{3+}$*
The total number of ligands around the central Al^{3+} ion is six, and so $x + y = 6$. The charge on the central ion is $3+$. This is not changed by the number of neutral water molecules. Each hydroxide ion has one negative charge, and so the

128

overall charge $z = (3 - y)$. If $y = 3$ the complex is uncharged, but if $y > 3$ it is negatively charged.

9 Answer **C**   *Knowledge of properties of aluminium halides*
Aluminium fluoride is largely ionic in character and does not react with water. Aluminium is able, by using d orbitals, to form complex ions with a co-ordination number of six, e.g. $[AlF_6]^{3-}$. This consists of a cental Al^{3+} ion surrounded by six fluoride ions. Aluminium bromide is largely covalent in character and will dissolve in some organic solvents. Aluminium chloride exists as a dimer, Al_2Cl_6, below 400 °C.

10 Answer **C**   *Comprehension of the hydrolysis of* $[Al(H_2O)_6]^{3+}$
The Al^{3+} ion has a large charge: radius ratio and is able to distort the O—H bond in co-ordinated water molecules. This results in cleavage of the O—H bond and the complex acts as an acid by releasing a proton. Addition of an acid moves the position of equilibrium to the left, suppressing hydrolysis. Addition of sodium carbonate, which removes protons, moves the position of equilibrium to the right.

11 Answer **D**   *Knowledge of the chemistry of aluminium chloride*
Aluminium chloride is a predominently covalent molecule. The small difference in electronegativity between Al and Cl, and the large energy needed to make Al^{3+} contribute to this. At room temperature the compound exists as the dimer Al_2Cl_6. Each aluminium is surrounded by four chlorine atoms. Above 800 °C the compound exists as single molecules, with a trigonal–planar arrangement of atoms.

12 Answer **B**   *Knowledge of the hydrolysis of* BCl_3
The hydrolysis of boron trichloride produces H_3BO_3, boric acid, as the product.

$$BCl_3 + 3 H_2O \rightarrow H_3BO_3 + 3 HCl$$

Boron is a non-metal and does not form the B^{3+} ion owing to the large amount of energy required to bring about ionization.

13 Answer **D**   *Knowledge of the flame test of magnesium*
In this respect, magnesium behaves in the same way as other substances in that electrons, falling from excited states to lower energy levels, give rise to the emission of radiation. This radiation, in the case of magnesium, is in the ultra-violet region of the electromagnetic spectrum and, therefore, cannot be seen by the eye.

14 Answer **B**   *Knowledge of electronic configuration of lithium*
The electronic configuration of lithium is $1s^2 2s^1$. Therefore, the maximum number of pairs of electrons that can be found in the atom is one.

                                                    Knowledge of reasons for similarity of
15 Answer **B**   *chemical properties within a group*
It is the presence of the lone electron outside a noble gas core which causes the similarities in chemical properties. Responses **C** and **D** are results of the presence of the lone s electron. These are symptoms of similar chemical properties not the reasons for them. Similarity of crystal structures in no way guarantees similarity of chemical properties.

16 **Answer C**   *Comprehension of reasons for the 'diagonal relationship'*
Lithium and magnesium would be expected to have different properties, as they
are in different groups. They have similar properties because of their similarity
in atomic radius and in electronegativity. Their electronegativities are low.

17 **Answer D**   *Application of knowledge of redox systems*
In the reaction hydride ions (H^-) are oxidised and protons (H^+) are reduced. The
oxidation states of calcium and oxygen remain unchanged and no proton
transfer occurs.

18 **Answer A**   *Knowledge of conditions necessary for alum formation*
An alum contains a singly charged cation and a trebly charged cation. Lithium
will not form an alum because its cation is too small to be stable within the crystal
structure.

19 **Answer A**   *Comprehension of the relationships between concentration,*
                        volume and formulae in ionic solutions
20 **Answer D**

21 **Answer C**
From a knowledge of the volume of a solution and its concentration it is possible
to work out the *number of moles of substance* present:

$$\text{No. of moles} = \text{volume (in } dm^3) \times \text{molarity.}$$

From a knowledge of the number of ions (cations and anions) present in 1
formula unit of the substance, e.g. $CaCl_2$ contains 3, it is possible to work out the
number of moles of ions present. When discussing concentrations of solutions,
the volume of the solution is irrelevant.

Solution	No. of moles of ions	Concentration of cations mol dm^{-3}	Concentration of anions mol dm^{-3}
A	8	2	2
B	3	1	2
C	3	0.5	1
D	7.5	5	2.5

XII Groups IV, V and VI

1 Answer **D**   *Knowledge of properties of Group(IV) halides.*
The tetrachloride of carbon is not hydrolyzed, unlike those of silicon, germanium, tin and lead. Carbon has a maximum covalency of four and has no d orbitals. It is believed that d orbitals are involved in the mechanism of hydrolysis of the other chlorides.

2 Answer **D**   *Knowledge of stable oxidation states of Group IV elements*
The most stable oxidation state of lead is $+2$. As Group IV is descended the $+4$ oxidation state becomes less stable and the $+2$ oxidation state becomes more stable. The $6s^2$ electrons in lead(II) compounds remain unused as an inert pair.

3 Answer **B**   *Knowledge of trends in Group IV elements*
The stability of oxidation state $+4$ decreases down the group, and so lead(IV) compounds are good oxidizing agents. Only tin and lead can form divalent cations, but complexes of the type $[MF_6]^{2-}$, involving d orbitals, are known for all Group IV elements except carbon.

                                  Knowledge of action of concentrated
4 Answer **D**   *nitric acid on Group IV elements*
Silicon, germanium and tin react with concentrated nitric acid according to the equation

$$3\,M + 4\,HNO_3 \;\rightarrow\; 3\,MO_2 + 4\,NO + 2\,H_2O.$$

Nitric acid is unable to oxidize lead beyond oxidation state $+2$ and lead(II) nitrate is the product.

5 Answer **A**   *Comprehension of reactions of tin(IV) chloride*
The product is the complex ion $[SnCl_6]^{2-}$. The complex is octahedral and tin is in the oxidation state $+4$.

6 Answer **D**   *Knowledge of bond types in Group IV halides*
The most ionic character results from bonding between elements with the greatest difference in electronegativity. Lead is the least electronegative Group IV element. Fluorine is the most electronegative halogen.

7 Answer **B**   *Knowledge of method of preparation of* $SiCl_4$
The process involves formation of silicon in oxidation state $+4$. When starting from silicon a strong oxidising agent is needed (chlorine). Although silicon is in oxidation state $+4$ in SiO_2 neither reaction **C** nor **D** is a way of removing oxygen.

8 Answer **B**   *Comprehension of the number of electrons in a molecular species*
The answer is found by adding the number of electrons for each atom and adjusting for negative/positive charge.

$$NO_3^{-} = 7 + 8 + 8 + 8 + 1 = 32 \qquad N_2O_4 = 7 + 7 + 8 + 8 + 8 + 8 = 46$$

$$NO_2 = 7 + 8 + 8 = 23 \qquad\qquad NO_2^{+} = 7 + 8 + 8 - 1 = 22$$

The 'odd electron' in NO_2 results in several interesting properties, e.g. it is paramagnetic, highly coloured and forms a dimer.

9 Answer **B**   *Knowledge of trends in Group V*
There is an increase in metallic character down the group, resulting in increasing basicity of the oxide and reduced thermal stability of the hydride. Only arsenic, antimony and bismuth are capable of forming ions of the type M^{3+} in solution.

10 Answer **C**   *Knowledge of trends in Group V hydrides*
Ammonia has a high boiling point owing to intermolecular attractive forces between the polar molecules. The strength of the dipole in phosphine is much less than that in ammonia and phosphine is less soluble in water. Phosphine is less thermally stable than ammonia and so it is a more powerful reducing agent.

11 Answer **C**   *Knowledge of reactions producing ammonia*
Warming an amide or an ammonium salt with a strong alkali will produce ammonia. Ammonia is produced when ionic nitrides react with water. Thermal decomposition of ammonium nitrite, however, yields nitrogen:

$$NH_4NO_2(s) \rightarrow 2 H_2O(g) + N_2(g).$$

12 Answer **A**   *Knowledge of the reactions of ammonia solution*
Addition of ammonia solution can cause the precipitation of many metal hydroxides. However, the precipitate may dissolve in excess ammonia solution if it is able to form a soluble ammine complex, e.g. $[Ag(NH_3)_2]^+$. The precipitate of $Fe(OH)_3$ is not dissolved in excess ammonia solution because there is no ammine complex of iron(III).

13 Answer **C**   *Knowledge of the shape of the ammonia molecule*
The molecule consists of three covalent $N—H$ bonds with one lone pair of electrons on the nitrogen atom. The repulsion between lone pair/bonding pairs is greater than that between bonding pairs and so the $H—N—H$ bond angle is reduced from the tetrahedral angle to about $107°$.

14 Answer **C**   *Comprehension of the oxidation states of nitrogen*
Nitrogen can exhibit oxidation states from $+5$ to -3. If an oxidation number $+1$ is assigned to hydrogen and -2 to oxygen the oxidation number (n) of nitrogen in NH_2OH can be found:

$$n + 2(+1) + (-2) + (+1) = 0$$

$$n = -1.$$

15 Answer **B**   *Comprehension of the redox reactions of hydrogen peroxide*
If hydrogen peroxide acts as a reducing agent it is itself oxidized to oxygen.

$$H_2O_2(aq) \rightarrow O_2(g) + 2 H^+(aq) + 2 e^-.$$

In the other responses hydrogen peroxide is reduced to water.

16 Answer **C**   *Comprehension of the redox reactions of hydrogen sulphide*
Only in reaction **C** is hydrogen sulphide reacting as a reducing agent. The sulphuric acid is reduced to sulphur dioxide and the hydrogen sulphide is oxidized to sulphur:

$$H_2S(aq) \rightarrow S(s) + 2 H^+(aq) + 2 e^-.$$

17 Answer **B**   *Knowledge of properties of Group VI hydrides*
In a series such as this it is expected that the molecule with the smallest molecular mass will have the lowest boiling point, i.e. H_2O. However, hydrogen bonding in water causes the boiling point to be higher than expected, as strong intermolecular bonds must be broken before vaporisation occurs.

18 Answer **A**   *Comprehension of the bonding in nitrogen compounds*
In its ground state nitrogen has a filled 2s orbital and three half-filled 2p orbitals. One dative bond can be formed by the donation of the 2s electrons and three normal covalent bonds can be formed by accepting three electrons from three other atoms.

19 Answer **D**   *Comprehension of the bonding in phosphorus compounds*
Promotion of one 3s electron will generate five half-filled orbitals and therefore five covalent bonds can be formed.

Comprehension of redox reactions of sulphur,
20 Answer **C**   *its compounds and its ions*
Redox reactions involve changes in oxidation number. In response **A** sulphur is oxidized from oxidation state $+4$ to $+6$. In response **B** the change is from 0 to $+2$ and in response **D** it is from $+2$ to $+3$. In response **C** both species have the same oxidation number $(+4)$.

XIII Group VII

1 Answer **D**   *Comprehension of trends within Group VII*
When descending Group VII, the atomic radius of the element increases and the
first ionization energy decreases. Electronegativity decreases down the group
and the radius of the negative ion will increase with the increasing radius of the
parent atom.

                              Knowledge of the formation and properties
2 Answer **B**   *of chlorine free radicals*
A chlorine free radical which is formed by the homolytic fission of a chlorine–
chlorine bond by exposure to ultra-violet radiation, is a chlorine atom with an
electron in an excited state. The free radical must contain the same number of
electrons (17) as the parent atom. Hydrogen is readily removed from hydrocarbons by chlorine free radicals forming hydrogen chloride. The combination
of chlorine free radicals gives rise to chlorine molecules:

$$Cl \cdot + Cl \cdot \rightarrow Cl_2$$

The excited electrons are used in forming the chlorine–chlorine bond.

3 Answer **D**   *Knowledge of the properties of Group VII elements*
The electrode potential for the fluorine/fluoride system is $+2.85$ V at $25\,^{\circ}C$.
The potentials for the other halogen/halide systems become increasingly negative as the group is descended. Consequently, the oxidizing power of the
halogens decreases as the group is descended and fluorine will oxidize all the
other halide ions.

4 Answer **A**   *Comprehension of the reactions of halogens*
Chlorine and bromine react with water as follows:

$$X_2 + H_2O \rightleftharpoons HX(aq) + HOX(aq).$$

Halogen oxidation numbers: 0 -1 $+1$

In this reaction disproportionation (the simultaneous oxidation and reduction of
an atomic species) occurs. The reaction of chlorine with potassium hydroxide
solution results in the formation of the products KCl(aq) and KOCl(aq) and this
reaction is also an example of disproportionation. Flurorine, however, is such a
powerful oxidizing agent that water is oxidized to oxygen and hydrogen fluoride
is formed. Fluorine is reduced from the zero oxidation state to oxidation state
-1.

5 Answer **A**   *Knowledge of the properties of Group VII hydrides*
In a series of similar molecules such as this, the compound with the highest
relative molecular mass is expected to have the highest boiling point. With
hydrogen fluoride, however, the large electronegativity of fluorine means that
the H—F bond is very polar and consequently there is a considerable degree of
hydrogen bonding between the molecules. This results in the boiling point of the
compound (293 K) being higher than that of hydrogen iodide (238 K).

6 Answer **D**   *Knowledge of the reactions of fluorine*
Fluorine is a powerful oxidizing agent and will combine with many elements forming compounds exhibiting the maximum oxidation state of the element. In CF_4, SF_6 and PF_5, however, the non-halogen element is already in its maximum oxidation state and cannot be further oxidized. The reaction with xenon to produce XeF_4 is well known.

7 Answer **A**   *Knowledge of trends in Group VII*
The heat of atomization of fluorine is extremely low (79 kJ mol^{-1}). Chlorine has a value of 121 kJ mol^{-1}. Fluorine atoms are very small and it is believed that when they approach closely in an F_2 molecule, there is repulsion between non-bonding electrons on the atoms creating a relatively weak bond.

8 Answer **B**
Comprehension of the steps in the formation of a hydrated chloride ion from chlorine
The conversion may be split into three steps:

$$\tfrac{1}{2} Cl_2(g) \rightarrow Cl(g) \text{ (enthalpy change of atomization)}$$

$$Cl(g) + e^- \rightarrow Cl^-(g) \text{ (electron affinity)}$$

$$Cl^-(g) \rightarrow Cl^-(aq) \text{ (hydration energy)}.$$

The ionization energy is *not* required.

9 Answer **A**   *Knowledge of silver(I) halides*
Silver chloride, silver bromide and silver iodide are all insoluble in water. Silver fluoride, however, is soluble because the low lattice energy which it possesses is overcome by the large hydration energy associated with the small fluoride ion.

10 Answer **C**   *Knowledge of the reactions of sodium halides*
Concentrated sulphuric acid will react with sodium halides according to the equation:

$$NaX(s) + H_2SO_4(l) \rightarrow NaHSO_4(s) + HX(g).$$

In the case of the bromide and iodide, however, the hydrogen bromide and hydrogen iodide formed are oxidized by the concentrated acid to yield the respective halogens. This is not so in the case of sodium chloride or sodium fluoride. With the chloride, however, if manganese(IV) oxide is present, the hydrogen chloride formed in the reaction is oxidized to chlorine.

11 Answer **B**   *Knowledge of the reactions of chlorine*
Chlorine cannot oxidize calcium chloride or silicon tetrachloride as the non-halogen element is in its maximum oxidation state. Chlorine will displace iodine from sodium iodide solution but will not displace fluorine from sodium fluoride solution.

12 Answer **B**   *Knowledge of the reactions of silver chloride*
Silver chloride will dissolve in ammonia solution to form the complex $[Ag(NH_3)_2]^+(aq)$. The chloride is virtually insoluble in the other substances.

13 Answer **B**   *Comprehension of oxidation number changes*
The equation for the reaction is:

$$I_2(aq) + 2 S_2O_3{}^{2-}(aq) \rightarrow S_4O_6{}^{2-}(aq) + 2 I^-(aq).$$

Iodine, oxidation number (0), is reduced to iodide (−1) and the sulphur in thiosulphate(VI) (+2), is oxidized to tetrathionate ($+2\frac{1}{2}$). One mole of iodine molecules undergoes a 2 electron reduction: $2 \times (0$ to $-1)$. One mole of thiosulphate(VI) undergoes a 1 electron oxidation: $2 \times (2$ to $2\frac{1}{2})$. The ratio $I_2 : S_2O_3{}^{2-}$, therefore, is $1 : 2$

14 Answer **D**

Knowledge of the standard enthalpy changes of formation of hydrogen halides

The standard enthalpy changes of formation of the hydrogen halides become more positive as the group is descended with that of hydrogen iodide being > 0. The result of this is that hydrogen iodide is the only hydrogen halide which can easily be decomposed into its elements. This can be done by putting a hot wire into the gas. The increasing size of the halogen as the group is descended results in a weakening of the hydrogen–halogen bond.

15 Answer **D**

Knowledge of the reactions of concentrated sulphuric acid with alkali metal halides

Concentrated sulphuric acid reacts with potassium chloride to generate hydrogen chloride gas. This gas is stable to further oxidation by the acid and no chlorine is formed unlike the results of similar reactions with potassium bromide and iodide.

16 Answer **A**   *Knowledge of the electronegativities of the halogens*

Electronegativities decrease as a group in the Periodic Table is descended. It follows that the halogen with the highest electronegativity is found at the top of the group. Electronegativity is a measure of the power of an atom to attract electrons to itself within a chemical bond.

17 Answer **C**   *Comprehension of the method of preparation of chlorine gas*

The chlorine is bubbled through water in order to remove any traces of hydrogen chloride (very soluble in water) and then passed through concentrated sulphuric acid to dry the gas.

XIV Transition Metal Chemistry

1 Answer C *Knowledge of factors affecting maximum oxidation number*
The maximum oxidation number of an element is controlled by the number of 4s and 3d electrons which can be removed from the atom. This rises to a maximum at manganese ($3d^5 4s^2$) and decreases thereafter, suggesting that the removal of paired 3d electrons is increasingly difficult.

2 Answer C *Comprehension of the oxidation states of atoms in compounds*
Stable complexes, with the oxidation state of the central atom equal to zero, are formed with only a few ligands. Carbon monoxide is one such example. In the other complexes, the oxidation states are Co(III), Fe(II) and Fe(III). This is found by knowing the charge on the ligand (H_2O and NH_3 are 0; OH^- is -1) and on the balancing ion (Cl^- is -1; SO_4^{2-} is -2).

3 Answer B *Knowledge of the oxidation of complex ions*
A stable complex ion of cobalt(III) is prepared by the oxidation of $[Co(NH_3)_6]^{2+}$. This takes place slowly in air. In the alternatives, the central metal ions exhibit the maximum oxidation number possible for each element.

4 Answer B *Knowledge of copper(I) chemistry*
The copper(I) ion is unstable in aqueous solution and it disproportionates into copper(II) ions and copper metal:

$$[Cu^{2+}(aq), Cu^+(aq)]\|Pt; \qquad E^{\ominus} = +0.15 \text{ V.}$$

$$Cu^+(aq)|Cu(s); \qquad E^{\ominus} = +0.52 \text{ V.}$$

Cu(II) ions will be present in the hydrated form: $[Cu(H_2O)_6]^{2+}$.

5 Answer D *Knowledge of chromium chemistry*
Chromate(VI) is prepared by the oxidation of a chromium(III) compound. The conversion of chromate(VI) into dichromate(VI) does not involve oxidation or reduction as there is no change in oxidation number. When dichromate(VI) is used as an oxidizing agent it is reduced to chromium(III). The common oxidation states of chromium are $+2$, $+3$ and $+6$.

6 Answer D *Comprehension of the meaning of the term 'co-ordination number'*
The co-ordination number of the metal in the complex is the number of monodentate ligands which are co-ordinated to it. In this example it is equal to six.

7 Answer B *Knowledge of cobalt complexes and their reactions*
The change is:

$$[Co(H_2O)_6]^{2+} \rightarrow [CoCl_4]^{2-}$$

octahedral	tetrahedral
pink	blue

In the presence of concentrated hydrochloric acid, the ligand water is replaced by chloride (Cl^-).

8 Answer **C**   *Knowledge of iron(II) chemistry*
   The change is:

$$[Fe(H_2O)_6]^{2+} \rightarrow [Fe(CN)_6]^{4-}.$$

The complex formed is octahedral (six ligands) and has an overall negative charge (anionic).

9 Answer **B**   *Knowledge of copper chemistry*
   The reaction of potassium iodide and copper(II) ions is unusual. Some of the iodide ions reduce copper(II) to copper(I) and are themselves oxidized to iodine (I_2).

$$2\,Cu^{2+}(aq) + 4\,I^-(aq) \rightarrow 2\,CuI(s) + I_2(aq).$$

The reaction is used to prepare CuI or to estimate the amount of copper(II) present by subsequent titration of the liberated iodine with thiosulphate(VI) solution.

                                    Comprehension of the changes in electronic
10 Answer **A**   *arrangements in transition metal ions*
   The electronic arrangement in a chromium atom is (Argon) $3d^5\,4s^1$. It is not (Argon) $3d^4\,4s^2$ because of the increased stability of a half-filled set of d orbitals. The 4s electron is lost first, followed by two 3d electrons, leaving the arrangement in Cr^{3+} as (Argon) $3d^3$.

                                    Comprehension of the changes occurring during the
11 Answer **C**   *preparation of manganate(VII) ions from manganese(IV) oxide*
   The changes are:
   (a) manganese(IV) oxide is oxidized to manganate(VI) ions:

$$4\,OH^-(aq) + MnO_2(s) \rightarrow MnO_4^{2-}(aq) + 2\,H_2O(l) + 2\,e^-$$
$$\quad\quad\quad\quad\text{black} \quad\quad\quad\quad \text{green}$$

   (b) oxidation of manganate(VI) ions to manganate(VII) ions:

$$MnO_4^{2-}(aq) \rightarrow MnO_4^-(aq) + e^-$$
$$\text{green} \quad\quad \text{purple}$$

12 Answer **C**   *Knowledge of copper chemistry*
   In the presence of a large concentration of chloride ions, the hexaaquacopper(II) ion is converted into the complex $[CuCl_4]^{2-}$. $[Cu(H_2O)_6]^{2+}$ only exists in aqueous solution. Crystallization of a solution of $[CuCl_4]^{2-}$ would yield $CuCl_2.2H_2O(s)$.

13 Answer **C**   *Comprehension of electronic arrangements in transition metals*
   Isoelectronic means having equal numbers of electrons, Vanadium has 23 electrons and so do Fe^{3+} $(26-3)$ and Mn^{2+} $(25-2)$. The Sc^{3+} ion has 18 electrons $(21-3)$.

14 Answer **B**   *Comprehension of geometrical isomerism*

A square planar complex $[MX_2Y_2]^{2-}$ may take the following forms:

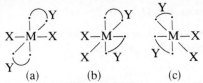

(a) is a cis complex with identical ligands adjacent.

(b) is a trans complex with identical ligands opposite.

15 Answer **D**   *Comprehension of geometric and optical isomerism*

The possible arrangements of the ligands about the metal ion are:

(a) is a trans complex and (b) and (c) are cis complexes. Arrangement (b), however, is the mirror image of (c) and the two isomers are non-superimposable and discrete. Therefore, each is optically active and solutions will rotate the plane of polarized light in opposite directions.

16 Answer **B**   *Knowledge of properties of ligands*

The ligand $H_2NCH_2CH_2NH_2$ can complex to a metal atom by donation of the lone pair of electrons on each nitrogen atom. It is described as a *bi*dentate ligand because it complexes at *two* sites.

17 Answer **C**   *Comprehension of the properties of complex ions*

Although there are three moles of chlorine per mole of complex, only two of those are present as ions. The third remains covalently bonded to the central chromium ion and consequently cannot be precipitated as silver chloride.

18 Answer **B**   *Comprehension of oxidation number changes*

The change is from Mn(VII) to Mn(V) and can be brought about in alkaline solution by the oxidation of the hydroxide ion:

$$2\,OH^-(aq) \;\rightarrow\; H_2O(l) + \tfrac{1}{2}O_2(g) + 2\,e^-.$$

Oxygen gas is evolved. The half-reaction for manganate(VII) is:

$$MnO_4^-(aq) + H_2O(l) + 2\,e^- \;\rightarrow\; MnO_3^-(aq) + 2\,OH^-(aq).$$

XV Fundamental Organic Chemistry

1 Answer **C** *Application of data to find molecular formulae*

The mass of one mole of gas = molar volume × density

$$= \quad 22.4 \quad \times \quad 2.5$$

$$= \quad 56\,g$$

The mass of one mole of C_4H_8 is $[4(12)+8(1)] = 56$ g.

2 Answer **B** *Application of data to determine the carbon content of a compound*

0.448 dm^3 of the compound represents

$$\frac{0.448}{22.4} = \frac{1}{50}\,\text{mole}.$$

If this produces 0.06 mole of carbon dioxide, then one mole will produce $50 \times 0.06 = 3$ mole of carbon dioxide. One molecule, therefore, will contain 3 atoms of carbon.

3 Answer **D** *Application of data to determine molecular formulae*

The relative molecular mass of the compound is twice the vapour density. For this compound it is $2 \times 56 = 112$, i.e. $(8 \times 12)+(16 \times 1)$. This corresponds to a molecular formula of C_8H_{16}.

4 Answer **A** *Application of data to determine empirical formulae*

0.112 dm^3 of the compound represents

$$\frac{0.112}{22.4} = \frac{1}{200}\,\text{mole}.$$

The mass of one mole, therefore, is $0.42 \times 200 = 84$ g. This corresponds to a hydrocarbon formula C_6H_{12}, i.e. $M_r = (6 \times 12)+(12 \times 1) = 84$. The empirical formula (simplest ratio of numbers of atoms), therefore, is CH_2. The formula C_2H_4 is *not* the simplest ratio.

5 Answer **A** *Knowledge of homolytic fission*

Homolytic fission is the breaking of a covalent bond in which one electron from the bonding pair returns to each of the previously bonded atoms. The electrons often return to excited states within the atom, forming free radicals. Common examples of homolytic fission are the formation of chlorine free radicals from chlorine molecules and the cracking of hydrocarbons. Responses **B**, **C** and **D** involve elimination or substitution reactions.

6 Answer **D** *Knowledge of the term 'nucleophile'*

A nucleophile is a negatively charged species, e.g. CN^-, or is a donor of electron pairs, e.g. NH_3 and H_2O. It seeks out a positively charged or electron deficient site. It cannot be positively charged itself. $NO_2{}^+$ is an electrophile.

7 Answer **C**   *Knowledge of bonding and hybridization*
The sp^2 hybrid bond is common in organic compounds. It is used in the formation of C=C bonds, e.g. in but-1-ene, and C=O bonds, e.g. in propanone and methanal. Bonding around the carbon atoms in ethanol involves sp^3 hybridization.

8 Answer **C**   *Comprehension of the cause of optical activity*
A molecule is optically active if four different atoms or groups of atoms are attached to the same carbon atom. This produces two isomers which are not superimposable on their mirror images. Alternative **C** is the only example with four different atoms/groups on one carbon atom, i.e. CH_3, H, OH and COOH.

Comprehension of the changes occurring during
9 Answer **C**   *the complete combustion of a hydrocarbon*
Complete combustion of a hydrocarbon converts all the carbon to carbon dioxide and all the hydrogen to water. One mole of oxygen (O_2) is required for each mole of carbon atoms (C) and 0.25 mole of oxygen is required for each mole of hydrogen atoms (H).

$$C_xH_y + \left(x + \frac{y}{4}\right) O_2 \rightarrow x\ CO_2 + \frac{y}{2} H_2O.$$

If $x = 5$ and $y = 12$,

$$x + \frac{y}{4} = 5 + \frac{12}{4} = 8$$

10 Answer **D**   *Comprehension of bond polarity*
A dipole arises in a bond if the atoms in the bond have different electronegativities. This occurs in the C—H bond in propane and the C—Cl bond in chlorobenzene. The C—C bond in 1,1-dichloroethane is polar because of the inductive effect. Chlorine atoms pull electrons towards themselves from the carbon atom inducing a dipole on the C—C bond. Ethane is a symmetrical molecule and there is no dipole on the central C—C bond.

Comprehension of the effect of the inductive
11 Answer **A**   *effect on acid strengths*
The weakest acid will have the lowest K_a value and the highest pK_a value. The strength of the acid depends on the degree of positive charge on the carboxyl carbon atom. The greater the positive charge, the greater the ease with which a proton is released. Propanoic acid is the weakest acid because the C_2H_5 group has the largest $+I$ effect, donating electrons to the carboyxl carbon atom. The CCl_3 group has the largest $-I$ effect, withdrawing electron from the carboxyl carbon atom, increasing the degree of positive charge on that carbon atom and facilitating, therefore, the loss of the proton.

12 Answer **C**   *Knowledge of organic acidity*
Phenol is a relatively weak acid ($pK_a = 10$). Both ethanoic and chloroethanoic acids (carboxylic acids) are stronger acids than phenol. Ethanoic acid is a weaker acid than methanoic acid because the CH_3 group has a larger electron donating effect ($+I$) than the single hydrogen atom. Trinitrophenol is a stronger acid than phenol because the nitro groups (NO_2) withdraw electrons from the ring and the O—H bond, promoting ionization and the release of the proton.

13 Answer **A** *Comprehension of the concept of a functional group*
The compound must either be an alcohol or an ether. These functional groups are $-OH$ and $-O-$ respectively and the only compounds that a molecular formula of C_2H_6O can represent are: CH_3CH_2OH and CH_3OCH_3, ethanol and ethoxyethane respectively. There is no way in which the other functional groups could be contained in compounds having that formula. The molecular formula of the nearest aldehyde is C_2H_4O, ethanal, and an ester must contain at least two oxygen atoms per molecule. An alkene does not contain oxygen.

14 Answer **B** *Comprehension of isomerism*
The formula C_4H_{10} can only represent the following compounds:

$CH_3CH_2CH_2CH_3$ (butane) and $CH_3CH(CH_3)CH_3$ (2-methylpropane).

15 Answer **B** *Comprehension of isomerism*
The following arrangements are possible:

$CH_3CH_2CH_2CH_2F$ (1-fluorobutane); $CH_3CH_2CH(F)CH_3$ (2-fluorobutane)

$CH_3CH(CH_3)CH_2F$ (1-fluoro-2-methylpropane)

$CH_3C(CH_3)(F)CH_3$ (2-fluoro-2-methylpropane).

16 Answer **C** *Comprehension of the relationship between empirical and molecular formulae*
The only possible compound is ethane, C_2H_6. Other multiples of CH_3 generate molecular formulae with too high a hydrogen : carbon ratio, e.g. C_3H_9.

17 Answer **D** *Comprehension of the calculation of empirical formulae*
Simplifying the ratios given, we obtain

$$C:H:O = 3.17:12.00:3.13$$

$$= 1.00:3.83:1.00$$

Converting to whole numbers, we find that

$$C:H:O = 1:4:1.$$

The empirical formula is CH_4O.
Clearly, in this example, the analysis figures are not precise and the value for hydrogen is too low.

18 Answer **B** *Knowledge of the structure of butan-2-ol*
The structure of butan-2-ol is $CH_3CH_2CH(OH)CH_3$:

$$
\begin{array}{cccc}
& H & H & OH\ H \\
& | & | & |\quad | \\
H- & C- & C- & C-\!\!-C-H \\
& | & | & |\quad | \\
& H & H & H\quad H
\end{array}
$$

There are nine C—H bonds.

19 Answer **D**   *Comprehension of bond angles in alcohols*
   The structure of methanol is:

$$\overset{\times\times}{\underset{H_3C}{}}\overset{\overset{\times\times}{O}\times}{\diagdown}\,_{H}$$

with two lone pairs of electrons on the oxygen atom. The bond angle (C—O—H)
in methanol is similar to the H—O—H angle in water, being slightly smaller than
the tetrahedral angle.

20 Answer **A**   *Comprehension of the structure of hydrocarbons*
   Hexane is a six carbon atom alkane (C_6H_{14}) with 14 C—H bonds. Only
   2,3-dimethylbutane, C_6H_{14}, will also contain 14 carbon–hydrogen bonds. The
   other responses involve alkanes with five or seven carbon atoms per molecule
   and correspondingly fewer or more carbon–hydrogen bonds than hexane.

143

XVI Aliphatic Hydrocarbons

1 Answer **D**   *Knowledge of IUPAC nomenclature*
A derivative of pentane must contain a chain of five carbon atoms. There are three methyl groups replacing hydrogen atoms on the chain and these are at positions 2 (two groups) and 3. The formula is:

$$H-\overset{\displaystyle H}{\underset{\displaystyle H}{C^1}}-\overset{\displaystyle CH_3}{\underset{\displaystyle CH_3}{C^2}}-\overset{\displaystyle H}{\underset{\displaystyle CH_3}{C^3}}-\overset{\displaystyle H}{\underset{\displaystyle H}{C^4}}-\overset{\displaystyle H}{\underset{\displaystyle H}{C^5}}-H.$$

This is written $CH_3C(CH_3)_2CH(CH_3)CH_2CH_3$.

2 Answer **C**   *Knowledge of the types of carbon–hydrogen bonds*
A tertiary C—H bond is one formed by a carbon atom which has no other hydrogen atoms bonded to it. There are usually three alkyl groups bonded to the carbon atom. 2-methylbutane is an example:

$$CH_3-CH_2-\overset{\displaystyle CH_3}{\underset{\displaystyle H}{C}}-CH_3$$

a tertiary C—H bond

3 Answer **C**   *Comprehension of the factors affecting boiling points in alkanes*
Boiling points of similar molecules generally increase as the relative molecular masses of the compounds increase. However, it is found that branched chain alkanes are more volatile than their unbranched isomers and the greater the degree of branching, the greater is the volatility. The boiling points of the compounds are: **A** 309 K; **B** 301 K; **C** 283 K; **D** 341 K. In alkanes, the bonding between molecules is the van der Waals type and the longer the molecule is, the greater is the contact between adjacent molecules and the more effective the bonding can be. When branching is introduced into the molecule, however, the degree of contact between the adjacent molecules is reduced and the bonding is consequently reduced. This results in the compound becoming more volatile. Response **C**, 2,2-dimethylpropane is a compact molecule which has relatively little contact with molecules alongside it. When considering melting points, however, the figures are: **A** 143 K; **B** 113 K; **C** 257 K; **D** 178 K. Here it is the approximately spherical nature of the molecule (**C**) which results in very close contact between molecules in the solid lattice and gives rise to a high melting point.

4 Answer **D**   *Knowledge of the reactions of chlorine and methane*
The reaction involves the homolytic fission of chlorine molecules producing free radicals. These are very reactive entities and remove a hydrogen atom from the methane molecule producing hydrogen chloride and a methyl free radical:

$$CH_4 + Cl^{\cdot} \rightarrow CH_3^{\cdot} + HCl.$$

The methyl free radical $CH_3\overset{\centerdot}{}$ can now react with a molecule of chlorine forming chloromethane and a chlorine free radical, which can go on to react further. A hydrogen atom can be abstracted from a molecule of chloromethane by a chlorine free radical to form the free radical $CH_2Cl\overset{\centerdot}{}$ which can then go on to react with another chlorine molecule, forming yet another chlorine free radical and a molecule of dichloromethane. The reaction can continue in this fashion forming trichloromethane and tetrachloromethane.

5 Answer **A**   *Comprehension of the effect of substituting atoms in a structure*
2,3-dimethylbutane has the structure

$$
\begin{array}{ccccccc}
& H & & CH_3 & H & & H \\
& | & & | & | & & | \\
H- & C^1 & - & C^2 & - C^3 & - & C^4 & -H. \\
& | & & | & | & & | \\
& H & & H & CH_3 & & H
\end{array}
$$

The methyl groups in positions two and three are equivalent because the molecule is symmetrical about the mid position of carbon atoms two and three. The two possible substitution products are:

(a)
$$
\begin{array}{ccccccc}
H & CH_3 & Cl & H \\
| & | & | & | \\
H-C-C & - & C & - & C-H \\
| & | & | & | \\
H & H & CH_3 & H
\end{array}
$$
and

(b)
$$
\begin{array}{ccccccc}
H & CH_3 & H & H \\
| & | & | & | \\
H-C-C & - & C & - & C-Cl. \\
| & | & | & | \\
H & H & CH_3 & H
\end{array}
$$

The structure

$$
\begin{array}{ccccccc}
H & CH_3 & H & H \\
| & | & | & | \\
H-C-C & - & C & - & C-H. \\
| & | & | & | \\
H & H & CH_2Cl & H
\end{array}
$$

is, of course, the same structure as (b).

6 Answer **C**
Comprehension of the bond breaking during the
ozonolysis of an alkene
Most alkenes react with ozone to form a cyclic peroxide 'oxonide':
e.g.

$$(CH_3)_2C{=}CH_2 + O_3 \rightarrow (CH_3)_2C\underset{O-O}{\overset{O}{\diagdown \diagup}}CH_2$$

These explosive compounds are not usually isolated but can be converted into many different products such as ketones, aldehydes, acids and alcohols. When treatment with ozone is used for identification purposes, the ozonide is usually hydrolyzed with a mixture of zinc metal and water. The products are aldehydes, ketones or a mixture of both. The composition of the mixture is dependent on the alkyl groups on the carbon atoms which have the double bond between them and, hence, the products of the hydrolysis can be used to identify the original alkene. If propanone and methanal are formed then the structure of the alkene is:

$$(CH_3)_2C{=}CH_2$$
$$\downarrow \qquad \downarrow$$
$$(CH_3)_2CO + CH_2O.$$

Ozonolysis is usually carried out by passing ozone-rich oxygen into a solution of the alkene in a solvent such as tetrachloromethane, which is stable to ozone.

7 Answer C   *Knowledge of a test for unsaturation*
The immediate decolorization of a solution of bromine in tetrachloromethane is a test for alkene unsaturation. Ethane (**A**) and hexane (**D**) are saturated hydrocarbons and do not give this reaction. Benzene does not contain 'alkene' double bonds and therefore will not react with bromine by simple addition, although the compound will react with bromine in the presence of a halogen carrier to form substitution products. But-1-ene will decolorize the bromine solution, reacting to form 1,2-dibromobutane.

8 Answer A   *Knowledge of IUPAC nomenclature*
The name tells us that there are five carbon atoms in a chain (pent) and that there is a double bond between carbon atoms number two and three (-2-ene). In addition, there is a methyl group substituted into the chain at position three and the skeleton of the molecule is

$$-\overset{|}{\underset{|}{C}}{}^1-\overset{|}{\underset{|}{C}}{}^2{=}\overset{|}{\underset{\underset{\displaystyle CH_3}{|}}{C}}{}^3-\overset{|}{\underset{|}{C}}{}^4-\overset{|}{\underset{|}{C}}{}^5-.$$

9 Answer B   *Application of Markownikoff's rule*
Addition of hydrogen bromide to an asymmetrical alkene results in the hydrogen atom of the hydrogen bromide molecule being bonded to the carbon atom of the double bond which already has the most hydrogens on it. This occurs because the proton is added to form the most stable carbonium ion. In this example, there are two possibilities: (i) $(CH_3)_2C^+CH_2CH_3$ and (ii) $(CH_3)_2CHCH^+CH_3$. Carbonium ion (i) is the more stable of the two because the (+I) effect of the three alkyl groups pushes electrons towards the positive charge, stabilizing it to a greater extent than in (ii).

The final step in the reaction is the acceptance of a Br^- ion by the carbonium ion to form product **B**:

$$\begin{array}{c} CH_3 \\ | \\ -\overset{|}{\underset{|}{C}}-\overset{|}{\underset{\underset{\displaystyle Br}{|}}{C}}-\overset{|}{\underset{|}{C}}-\overset{|}{\underset{|}{C}} \end{array}$$

10 Answer D   *Knowledge of electrophilic addition to an alkene*
In bromine water the reactive species towards double bonds is the polar bromic(I) acid,

$$\overset{\delta+}{Br}{-}\overset{\delta-}{OH}.$$

Reaction occurs in the following manner:

$$\underset{\underset{\displaystyle Br\frown OH}{}}{C=C} \quad \rightarrow \quad -\overset{|}{\underset{|}{C}}-\overset{+}{\underset{|}{C}}- \ + OH^-.$$

11 Answer B   *Knowledge of electrophilic addition to an alkene*
Addition of sulphuric acid proceeds via proton addition to produce a stable secondary carbonium ion: $CH_3CH_2CH^+CH_3$. The hydrogensulphate group then becomes bonded to position 2 in the chain. Hydrolysis then replaces the HSO_4^- group by an OH group and butan-2-ol is formed: $CH_3CH_2CH(OH)CH_3$. Note that the carbonium ion is stabilized by the presence of the ethyl group and the methyl group. Had the positive charge been on the carbon atom at the end of the chain, there would only have been one alkyl group to effect this stabilization of the primary carbonium ion.

12 Answer A   *Comprehension of the stability of carbonium ions*
The order for the stability of carbonium ions is: tertiary > secondary > primary, because there is increasing stabilization of the positive charge on the ion by alkyl groups as they push electrons on to the carbon atom nominally carrying the positive charge. See questions nine and eleven.

13 Answer C   *Application of knowledge to experimental data*
The reaction is:

$$C_3H_4(g) + 2\,H_2(g) \ \rightarrow \ C_3H_8(g).$$

Propyne contains a carbon–carbon triple bond. One mole of propyne requires two moles of hydrogen molecules to bring about saturation. Therefore, 6 dm^3 of hydrogen will react with 3 dm^3 of propyne.

14 Answer D   *Knowledge of the oxidation reactions of alkenes*
This reaction is a conversion of an alkene to an epoxyalkane. Peroxobenzoic acid is a suitable reagent here, although peroxotrifluoroethanoic acid, CF_3COOOH, is also effective. Ozone produces carbonyl compounds (question six), potassium manganate(VII) generates a diol and concentrated sulphuric acid/hydrolysis produces alcohols.

15 Answer D   *Knowledge of the properties of hydrocarbons*
A sooty flame contains unoxidized carbon particles, indicating the incomplete combustion of a substance. Although other factors may be involved, in a simple comparative test such as this one, the higher the carbon : hydrogen ratio, the smaller is the proportion of carbon that is oxidized and the sootier is the flame. The simple ratios (empirical formulae) in these cases are:

     A CH_4;   **B** CH_3;   **C** CH_2;   **D** CH.

16 Answer C   *Comprehension of the reaction between soda lime and sodium ethanoate*

The equation for the reaction is:

$$CH_3COONa + NaOH \ \rightarrow \ CH_4 + Na_2CO_3.$$

This reaction is only suitable for use with sodium ethanoate in the production of methane. Higher members of the series, e.g. propanoates decompose to produce a mixture of alkanes, unsaturated compounds and hydrogen.

17 Answer **B**   *Application of knowledge to the structure of aldehydes and ketones*

The equation for this reaction is:

$$2C_2H_4 + O_2 \rightarrow 2\ CH_3-CHO.$$

It can be seen that 0.5 mole of oxygen molecules are required per mole of ethene.

18 Answer **C**   *Knowledge of the reactions of ethyne*

Generally, alkynes exhibit similar properties to alkenes and so will decolorize acidified manganate(VII) solution. They are oxidized by manganate(VII). Addition of hydrogen to ethene will generate ethane, C_2H_6. When ethyne is burnt completely, three moles of products are formed per mole of ethyne:

$$C_2H_2 + 2\tfrac{1}{2}O_2 \rightarrow 2\ CO_2 + H_2O.$$

The brown powder, Cu_2C_2, formed when ethyne reacts with copper(I) chloride is explosive when dry, as is the corresponding silver salt. It contains the C_2^{2-} ion.

19 Answer **B**   *Comprehension of molecular structure*

The compound must contain three bonds. There is only one way in which this can be realized:

$$\diagdown\hspace{-0.3em}\diagup C = C - C \equiv C - \quad \text{(but-1-ene-3-yne)}$$

XVII Alcohols, Amines, Ethers and Haloalkanes

1 Answer **B**

Knowledge of the factors affecting the rate of hydrolysis of haloalkanes

The progress of the hydrolysis of the haloalkane can be followed by the formation of the precipitate of silver halide. The iodine compound reacts fastest because of the weakness of the C—I bond. This bond is weaker and longer than the corresponding C—Cl bond. The benzene derivatives are very difficult to hydrolyze because the carbon—halogen bond is strengthened by the overlap of a p orbital from the halogen with the π orbitals of the benzene ring.

2 Answer **C**

Knowledge of the factors affecting the rate of hydrolysis of haloalkanes

2-bromo-2-methylpropane is a tertiary haloalkane:

$$
\begin{array}{ccccc}
 & H & Br & H & \\
 & | & | & | & \\
H- & C- & C- & C- & H. \\
 & | & | & | & \\
 & H & | & H & \\
 & & H-C-H & & \\
 & & | & & \\
 & & H & & \\
\end{array}
$$

The bromine atom is bonded to a carbon atom which has three alkyl groups attached. The mechanism of hydrolysis for a tertiary alkane is via a carbonium ion. Tertiary carbonium ions are very stable compared to secondary and primary carbonium ions and are readily formed in aqueous solution.

3 Answer **D** *Knowledge of synthetic routes from haloalkanes*

The reaction steps are listed below:

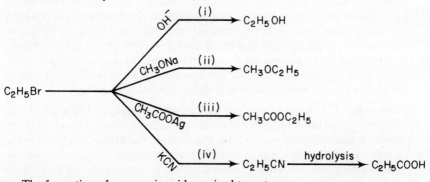

The formation of propanoic acid required two steps.

4 Answer **D**

Knowledge of the concurrent substitution and elimination reactions of haloalkanes

OH^- acts as a nucleophile in the substitution reaction with the OH group replacing the Br atom. Propene ($CH_3CH=CH_2$) is formed by the removal of a proton by OH^- (acting as a base) from the intermediate carbonium ion

$CH_3CH^+CH_3$. The use of alcohol as solvent favours the elimination reaction producing propene.

5 Answer **B**   *Comprehension of the meaning of the term 'tertiary alcohol'*
A tertiary alcohol is a compound whose OH group is attached to a carbon atom which has three alkyl groups attached.
e.g.

$$\begin{array}{ccc} & \text{H} & \text{OH} & \text{H} \\ & | & | & | \\ \text{H}-\text{C}-\text{C}&-&-\text{C}-\text{H} \\ & | & | & | \\ & \text{H} & & \text{H} \end{array}$$ (2-methylpropan-2-ol).

$$\begin{array}{c} \text{H}-\text{C}--\text{H} \\ | \\ \text{H} \end{array}$$

Pentan-3-ol and propan-2-ol are secondary in that they have the structure $>$CH(OH) within the molecule. Butan-1-ol is a primary alcohol ($-$CH$_2$OH). It is important to realize that the numbers appearing in the IUPAC nomenclature do not correspond to any particular type of alcohol.

6 Answer **B**   *Knowledge of the general properties of alcohols*
Inter-molecular hydrogen bonding is found in alcohols. The O$-$H bond in alcohols is polar and results in hydrogen bonding *between* adjacent molecules. *Intra*-molecular hydrogen bonding occurs within a single molecule, an example being the benzene derivative possessing an OH and NO$_2$ group: 2-nitrophenol.

the intermolecular
← hydrogen bond

Alcohols can behave as acids in that they can donate protons to bases. They are, however, weaker acids than water unless there is an electron-withdrawing substituent within the molecule. This can pull the electron pair away from the hydrogen atom in the O$-$H bond by a $-$I inductive effect, facilitating the release of a proton. The pK_a value of ethanol is 15.5 whereas that of 2,2,2-trifluoro-ethanol is 12.4. It is possible for an alcohol to act as a base by accepting a proton through bonding with a lone pair on the oxygen atom. Reaction with sodium produces an ionic sodium salt, e.g. sodium ethoxide, C_2H_5ONa.

7 Answer **B**   *Knowledge of the mechanism of the dehydration of an alcohol*
The first step in the dehydration of ethanol is the protonation of the molecule by concentrated sulphuric acid to form $C_2H_5OH_2^+$. The proton is bonded to the alcohol through a lone pair of electrons on the oxygen atom. Subsequent to the first step, the ion loses a molecule of water forming the carbonium ion $C_2H_5^+$, which then forms ethene by the loss of a proton.

8 Answer **C**   *Knowledge of oxidation reactions of alcohols*
Primary alcohols are easily oxidized to aldehydes by warming with acidified dichromate(VI) solution. Further oxidation to an acid is effected by using a more concentrated acidic solution. In a similar fashion to the oxidation of the primary alcohol to an aldehyde, a secondary alcohol can be oxidized to a ketone. Tertiary

alcohols are very difficult to oxidize because oxidation involves the rupture of a carbon–carbon bond and this needs very vigorous conditions. Of the alternatives listed, only $(CH_3)_3COH$ is a tertiary alcohol, the others being primary or secondary.

9 Answer **C**   *Knowledge of the oxidation of alcohols*
Oxidation of a secondary alcohol produces a ketone. Pentan-2-one is produced by the oxidation of the alcohol possessing an OH group on carbon atom number 2. This alcohol is pentan-2-ol.

10 Answer **C**   *Knowledge of the triiodomethane (iodoform) reaction*
This reaction may be used to test for alcohols containing the $CH_3CH(OH)$ group. In the test this group is oxidized to a ketone group, CH_3CO, which can then be converted to the yellow triiodomethane $CHI_3(s)$. Of the alternatives given only propan-2-ol contains this grouping. The test is also given by ethanol.

11 Answer **D**   *Knowledge of the reactions of alcohols*
An alcohol will form an ester by the reaction with a carboxylic acid (in the presence of a strong acid catalyst), an acid anhydride or an acid chloride.

12 Answer **B**   *Comprehension of the methods of preparing ethers*
Ethers can be prepared by the nucleophilic substitution of a haloalkane with a sodium alkoxide. For example, 1-iodopropane and sodium ethoxide are mixed together and the ethereal product, 1-ethoxypropane is distilled off. The other product is sodium iodide. The use of 2-iodopropane would lead to 2-ethoxypropane.

$(CH_3)_2CHOC_2H_5$ $CH_3CH_2CH_2OC_2H_5$.
    2-ethoxypropane   1-ethoxypropane

13 Answer **B**   *Knowledge of the reactions of ethers*
Fission of a carbon–oxygen bond in an ether only occurs under vigorous conditions. In the reaction with hydrogen bromide an intermediate, $C_2H_5OH^+C_2H_5$, is formed. This is attacked by Br^- to yield ethanol and bromoethane.

14 Answer **B**   *Comprehension of mechanism type*
The reaction of bromoethane with OH^- ions involves nucleophilic attack by OH^- ions resulting in the substitution of Br by OH. The reaction between ethene and hydrogen bromide involves electrophilic attack (addition) and the reaction between ethanol and concentrated sulphuric acid is an elimination reaction. The reaction between potassium manganate(VII) solution and ethanol involves oxidation.

15 Answer **B**   *Knowledge of the reactions of alcohols and haloalkanes*

16 Answer **C**

17 Answer **C**

18 Answer **D**

19 Answer **C**

Questions sixteen, seventeen and nineteen involve elimination reactions, with the formation of alkenes. Question fifteen involves oxidation as does question eighteen.

20 Answer **A**   *Comprehension of the oxidation of alcohols*

The only two possible oxidation products of a primary alcohol whose formula is RCH_2OH are RCHO and RCOOH. The former involves the loss of two hydrogen atoms per molecule and the latter the loss of two hydrogen atoms and the gain of one oxygen atom. In the case quoted in the question, the only possible products are $C_6H_{10}O$ and $C_6H_{10}O_2$. Note that the alkyl group C_6H_{11} is unsaturated.

21 Answer **C**   *Comprehension of metamerism*

Ether **C**, 1-ethoxypropane, has the same molecular formula as:
(a) 2-methoxybutane: $CH_3OCH(CH_3)C_2H_5$.
(b) 1-methoxybutane: $CH_3OCH_2CH_2CH_2CH_3$.
(c) 2-ethoxypropane: $C_2H_5OCH(CH_3)_2$.
(d) 1-methoxy-2-methylpropane: $CH_3OCH_2CH(CH_3)CH_3$.
The other ethers in the question cannot have other ethers as structural isomers, as more than three carbon atoms must be present in the molecule before this can occur.

22 Answer **C**   *Knowledge of the meaning of the term 'tertiary amine'*

C is a tertiary amine because the nitrogen atom is not directly bonded to a hydrogen atom but is bonded to three carbon atoms. **A** is a primary amine ($-NH_2$), whilst **B** and **D** are secondary amines ($-NH$).

23 Answer **D**   *Knowledge of the methods of preparation of amines*

The Hofmann degradation of ethanamide (CH_3CONH_2) and the reduction of hydrogen cyanide both yield the primary amine methylamine (CH_3NH_2). The reductive amination of ethanal produces ethylamine ($C_2H_5NH_2$). Treatment of bromoethane with alcoholic ammonia solution produces a mixture of amines ($C_2H_5NH_2$, $(C_2H_5)_2NH$, $(C_2H_5)_3N$) and the quaternary ammonium compound $(C_2H_5)_4N^+Br^-$. The second of the amines listed is a secondary amine.

24 Answer **A**   *Knowledge of the basicity of amines*

A large pK_b value for a base is consistent with a low K_b value, i.e. the compound has a low base strength. Aliphatic amines are stronger bases than ammonia because the electron donating alkyl groups (+I) assist the donation of a lone pair of electrons on the nitrogen atom to a proton.

25 Answer **A**   *Knowledge of the reactions of amines*

A primary aliphatic amine forms an unstable diazonium salt with nitrous acid, decomposing immediately to give nitrogen. A primary aromatic amine forms a stable diazonium compound at 0 °C because of the stabilizing influence of the benzene ring. Second amines yield a nitrosoamine (yellow oil) under these conditions, and tertiary aromatic amines form salts.

XVIII Carboxylic Acids and their Derivatives

1 Answer **D**   *Knowledge of the methods of preparation of acid anhydrides*
Acid anhydrides can be made by distilling mixtures of an acid chloride and the
sodium salt of a carboxylic acid ($CH_3COCl + CH_3COONa$). A carboxylic acid is
not usually dehydrated by concentrated sulphuric acid, but acid anhydrides can
be prepared from higher members of the acid series by heating and fractionating
a mixture of the acid and ethanoic anhydride. The acid employed must have a
higher boiling point than ethanoic acid:

$$2\ RCOOH + (CH_3CO)_2O \rightarrow (RCO)_2O + 2\ CH_3COOH.$$

Refluxing silver ethanoate and a haloalkane in ethanol will produce an ester, as
will the reaction between ethanoic acid and ethanol in the presence of a strong
acid catalyst.

2 Answer **D**   *Knowledge of the bonding in amides*
Ethanamide, an acid amide, contains a carbonyl group ($C{=}O$) and the bonding
in this group is based upon sp^2 hybridized orbitals giving rise to approximate
bond angles of $120°$ about the carbon atom. The exact bond angle will depend on
the relative magnitudes of the other groups bonded to the carbon atom. In this
case these groups are NH_2 and CH_3.

$$\begin{array}{c} CH_3 \\ \diagdown \\ C{=}O. \\ \diagup \\ H_2N \end{array}$$

3 Answer **C**   *Knowledge of IUPAC nomenclature*
Under the IUPAC rules, the numbering in a carboxylic acid begins at the
COOH carbon, i.e.

$$CH_3 \ldots CH_2CH_2CH_2COOH$$

$$n \qquad 4 \quad 3 \quad 2 \quad 1$$

3-methylbutanoic acid consists of a four atom carbon chain, the first carbon
atom of which is the carboxylic acid carbon. Substituted on to the third carbon
atom of the chain there is a methyl group:

$$\begin{array}{c} ||| \\ -C-C-C-COOH. \\ ||| \\ CH_3 \end{array}$$

The other bonds from the carbon atom are to hydrogen atoms. This formula may
be written: $CH_3CH(CH_3)CH_2COOH$.

4 Answer **C**   *Knowledge of the reduction of organic compounds*
Carboxylic acids are not reduced by $NaBH_4$ in ethanol although they are
reduced by a mixture of $NaBH_4$ and $AlCl_3$ in diglyme, producing a primary
alcohol. They are reduced by $LiAlH_4$ in ether, again to a primary alcohol. Acid

153

anhydrides can also be reduced by $LiAlH_4$ in ether to primary alcohols. Ketones (**C**) are reduced to secondary alcohols by many reducing systems including $LiAlH_4$ in ether, $NaBH_4$ in ethanol, and B_2H_6 in tetrahydrofuran. Methanoic acid is dehydrated by concentrated sulphuric acid, forming carbon monoxide.

5 Answer **D**   *Knowledge of acid strengths and the hydrolysis of acid chlorides*
All four compounds would give rise to different degrees of acidity. Responses **A**, **B** and **C**, however, all give rise to relatively high pH values in solution because of incomplete dissociation of the molecule. Addition of ethanoyl chloride to water yields both the weak acid, ethanoic acid, and hydrogen chloride. This dissolves readily in the water giving a high concentration of hydrogen ions in the solution and resulting in a low pH value.

6 Answer **B**   *Knowledge of the preparation of acid amides*
Concentrated ammonia solution will react with propanoic acid to generate the ionic ammonium salt: $CH_3CH_2COO^-NH_4^+$. If this is heated the acid amide and water are produced. The other reactions produce the acid amide by a reaction which is thought to be a nucleophilic attack by ammonia on the carbon atom of the carbonyl group.

                        Comprehension of data concerning the hydrolysis
7 Answer **A**   *and formation of an ester*
In the reaction from right to left the ^{18}O atom in the alcohol molecule can be seen to be incorporated into the ester molecule. In the formation of the ester, it is the O—H bond in the alcohol which is broken and the C—O single bond which is broken in the acid.

8 Answer **D**   *Knowledge of the preparation of esters and their nomenclature*
Each of the pairs of reactants listed, under suitable conditions, would produce an ester. Alcohols react rapidly with acid anhydrides and acid chlorides to produce esters, although the rate varies with the nature of the alcohol. Tertiary alcohols, for example, react very slowly and often give rise to side reactions generating alkenes or haloalkanes. The reaction between an alcohol and a carboxylic acid requires a strong acid catalyst to produce a reasonable rate of reaction. The *ethyl* part of this ester must come from the alcohol reactant and the *propanoate* part must come from the carboxylic acid. Reactions **A**, **B** and **C** would produce propyl ethanoate.

9 Answer **C**   *Application of knowledge to the hydrolysis of acid chlorides*
When one mole of ethanoyl chloride is completely hydrolyzed, one mole of ethanoic acid is produced together with one mole of hydrogen chloride:

$$CH_3COCl + H_2O \rightarrow CH_3COOH + HCl.$$

0.2 mole of acid chloride, therefore, will generate 0.4 mole of monobasic acid. This will require 0.4 mole of sodium hydroxide to neutralize it. Therefore, 200 cm^3 of 2 M alkali are required.

10 Answer **A**   *Knowledge of the relative acidities of carboxylic acids*
The K_a values (mol dm^{-3}) are:

ethanoic acid	1.8×10^{-5}	(**A**).
benzoic acid	6.4×10^{-5}	(**B**).
chloroethanoic acid	1.6×10^{-3}	(**C**).
iodoethanoic acid	7.5×10^{-4}	(**D**).

Acids **B**, **C** and **D** are all stronger acids than ethanoic because the group R on the molecule, RCOOH, in each case is more effective in withdrawing electrons from the COOH group (−I effect) than is the methyl group in ethanoic acid. The greater the withdrawal of electrons from the COOH group, the lower is the electron density in the O—H bond, and the more easily is the proton lost:

$$\twoheadleftarrow C \twoheadleftarrow O \leftarrow H.$$
$$\parallel$$
$$O$$

11 Answer **C**   *Comprehension of the reactions of acid chlorides and acid anhydrides*
The N-alkyl acid amide, $CH_3CONH(C_3H_7)$, can be prepared only by reaction of propylamine ($C_3H_7NH_2$) with either ethanoyl chloride or ethanoic anhydride. The formulae of the products of reactions **A**, **B** and **D** are:

A $CH_3CONH(C_2H_5) + CH_3COOH$.
B $C_2H_5CONH(C_2H_5) + HCl$.
D $C_2H_5CONH(CH_3) + C_2H_5COOH$.

12 Answer **C**   *Knowledge of the Rosenmund reduction*
To reduce an acid chloride to an aldehyde without further reduction to the primary alcohol requires a relatively mild reductant and one which can overcome the problem that aldehydes are generally more easily reduced than acid chlorides. A catalyst of palladium which is poisoned with barium sulphate is used. The reaction is carried out in boiling dimethylbenzene and often other poisons such as quinoline or sulphur are added.

13 Answer **B**   *Knowledge of bonding in acids and their derivatives*
In non-aqueous solution carboxylic acids dimerize by hydrogen bonding between molecules. This gives rise to a higher than expected boiling point because the hydrogen bonds must be broken in order that the molecule can enter the gas phase. The stable ester, methyl ethanoate, does not possess a sufficiently positively charged hydrogen atom to take part in hydrogen bonding. The molar masses of the two isomers are, of course, equal.

14 Answer **A**   *Knowledge of the reduction of carboxylic acids*
Of the reductants named in the question, only lithium tetra-hydridoaluminate(III) can effect this change, although hydrogen and a ruthenium/charcoal catalyst or copper/barium/chromium oxide catalyst can be used. The acid may be reduced with a wider variety of reductants if it is first esterified or converted to the acid chloride.

15 Answer **D**   *Knowledge of the methods of preparation of acid anhydrides*
The reaction proceeds via a nucleophilic attack on the carbonyl carbon atom of the acid chloride by the carboxylate ion:

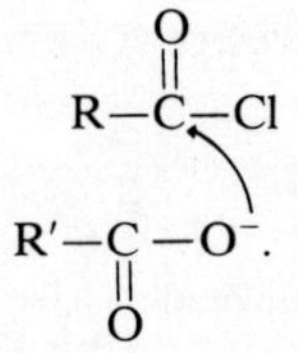

$$\text{R}-\overset{\displaystyle\overset{\text{O}}{\|}}{\text{C}}-\text{Cl}$$

$$\text{R}'-\overset{}{\underset{\displaystyle\underset{\text{O}}{\|}}{\text{C}}}-\text{O}^-.$$

The chlorine is lost as the negative ion and the acid anhydride molecule results.

Knowledge of the conditions and reagents
16 Answer **E**   *necessary for certain reactions to occur*

17 Answer **B**

18 Answer **C**

19 Answer **A**

Step **E** is an esterification and requires a strong acid catalyst. Heating the ionic ammonium salt of a carboxylic acid generates the amide. This can also be produced from the acid chloride by reaction with ammonia; hydrogen chloride is produced as a by-product in this reaction.

XIX Aldehydes and Ketones

1 Answer **B**   *Comprehension of molecular geometry*
The formulae of the compounds are: **A** CH_3OH; **B** $HCHO$; **C** $(CH_3)_2CO$; **D** $HCOOH$. Compounds containing a methyl group cannot be planar as this group involves a tetrahedral distribution of bonds around the carbon atom (sp^3 hybridization) and consequently must lie in more than one plane. A carboxylic acid group,

$$-C\diagup\!\!\!\!\diagdown\begin{array}{l} O \\ O-H \end{array}$$

involves an O—H bond which is not in the same plane as the other bonds in the group, because the presence of the two lone pairs on the oxygen atom of the OH group produces a tetrahedral arrangement of electron pairs about the oxygen atom; only two of the electron pairs constitute bonds. In methanal, there is a carbonyl group with two hydrogen atoms attached to the carbon atom. This involves sp^2 hybridization, with a planar molecule provided that single atoms are attached to the carbon of the carbonyl group.

2 Answer **C**   *Comprehension of IUPAC nomenclature*
The compound is a ketone (-one), involves a phenyl group, C_6H_5-, and possesses two other carbon atoms (-ethan-), including the carbonyl carbon atom.

3 Answer **D**   *Knowledge of addition to C=C and C=O bonds*
The C=O bond is more polar than the C=C bond because of the electronegativity differences between the atoms. Attack at the C=C bond is by an electrophile seeking out the electron-rich double bond. Addition of HCN to the C=C bond proceeds via a nucleophilic attack by CN^- on the positively polarized carbonyl carbon atom. However, if strongly acidic conditions prevail, this route is stopped by the protonation of the carbonyl group. The addition of HCN to carbonyl compounds is known as the cyanohydrin reaction and can be carried out for all aldehydes but the only common ketones that will undergo the reaction are propanone, butanone and pentan-3-one. Addition of HBr to the electron-rich C=C bond in alkenes proceeds by electrophilic attack by H^+. HCN is a relatively weak acid and provides too low a concentration of protons for this mode of attack to be successful.

4 Answer **C**   *Knowledge of the oxidation of aldehydes and ketones*
Aldehydes are easily oxidized to carboxylic acids by such reagents as acidified dichromate(VI), Fehling's solution and Tollen's reagent. Primary alcohols such as ethanol are also readily oxidized, firstly to aldehydes and then to acids. With ketones, however, because of the absence of a hydrogen atom on the carbonyl carbon atom, oxidation to a carboxylic acid containing the same number of carbon atoms as the ketone is impossible. Oxidation, however, can occur.

Relatively strong oxidants bring about rupture of a carbon–carbon bond forming two carboxylic acids:

$$CH_3COCH_2CH_3 \xrightarrow{\text{HNO}_3} 2\,CH_3COOH.$$

Note that ketones with the structure

are as easy to oxidize as aldehydes.

5 Answer **D**

Knowledge of the differences in reactions between aldehydes and ketones

Aldehydes and ketones form coloured, crystalline condensation products with 2,4-dinitrophenylhydrazine, e.g.

As we have seen in question four, only aldehydes can react with Tollen's reagent (ammoniacal silver nitrate solution) forming metallic silver and the oxidation product of the aldehyde. Ethanol does not react with either reagent.

6 Answer **C** *Comprehension of a condensation reaction of ketones*

The reaction proceeds via nucleophilic attack upon the carbonyl carbon atom by the lone pair of electrons on the NH_2 group forming an intermediate compound which then rearranges, losing water, to form propanone oxime:

$$\rightarrow (CH_3)_2C{=}NOH + H_2O.$$

7 Answer **B** *Knowledge of the triiodoform reaction*

This reaction is given by most compounds which contain the

and those compounds such as ethanol and propan-2-ol which can be oxidized under the conditions of the reaction to form this group. Of the four compounds shown, only propanal does not contain this group: The formula of propanal is CH_3CH_2CHO.

8 Answer **D** *Comprehension of addition/elimination reactions*

The reaction of ethanal with acidified manganate(VII) involves oxidation, whilst that of propanone with an ethereal suspension of $LiAlH_4$ involves reduction.

The reaction of an aldehyde with sodium hydrogensulphite is an addition reaction in that SO_3Na and H are added across the carbonyl double bond. This addition across the double bond also occurs with NH_2OH and butanone. Here H and NHOH are added across the bond forming

$$R \underset{R}{\overset{}{\diagup}} \overset{OH}{\underset{NHOH}{C}}$$

A molecule of water is then eliminated from this addition compound to form:

$$\underset{R}{\overset{R}{\diagdown}} C=N-OH,$$

an oxime. Oximes are well defined, crystalline solids which may be used to characterize carbonyl compounds.

9 Answer **C**   *Knowledge of the reductions carried out with LiAlH$_4$*
The four compound types shown are acid anhydride, acid chloride, ketone and ester. $LiAlH_4$ reduces the ketone to a secondary alcohol, $-CH(OH)$, and the others to primary alcohols, $-CH_2OH$.

Comprehension of the changes which occur
10 Answer **B**   *during the reactions of ketones*
Phenylhydrazine,

reacts with ketones by addition/elimination to form phenylhydrazones: $C_6H_5NHN=C(R)_2$. The number of extra carbon atoms here is six. Hydrogen cyanide undergoes the cyanohydrin reaction (question three) to form the addition product

$$R \underset{R}{\overset{}{\diagup}} \overset{OH}{\underset{CN}{C}}$$

which can be hydrolyzed to the acid.

$$R \underset{R}{\overset{}{\diagup}} \overset{OH}{\underset{COOH}{C}}$$

This process involves the addition of one extra carbon atom to the ketone. The aldol condensation involves the formation of a molecule containing a $C=O$ group and an $O-H$ group:

$$2(CH_3)_2CO \rightarrow (CH_3)_2C(OH)CH_2COCH_3.$$

159

Two ketone molecules join together, thereby increasing the number of carbon atoms in the molecule by a factor of two.

Most aldehydes will form resinous products when warmed with concentrated sodium hydroxide solution, but ketones do not show this reaction. Again with aldehydes, the Cannizzaro reaction can occur in 50% alkali solutions (question thirteen).

11 Answer **B**   *Comprehension of the formation of Schiff bases*
This is another example of an addition/elimination reaction involving amines and carbonyl compounds. The base $RCH = NR'$ is formed from the amine $R'NH_2$ and the aldehyde RCHO. The addition compound is

$$\begin{array}{ccc} H & & OH \\ & \diagdown \diagup & \\ & C & \\ & \diagup \diagdown & \\ R & & NHR' \end{array}$$

and this loses water to form

$$\begin{array}{c} R \\ \diagdown \\ \quad C{=}N{-}R'. \\ \diagup \\ H \end{array}$$

Clearly, the amine must be a primary amine because it loses one hydrogen atom during the addition reaction and one during the elimination reaction. The carbonyl compound must be an aldehyde because the base has an RCH structure which a ketone could not generate.

                                          Comprehension of the conditions necessary
12 Answer **B**   *for the aldol condensation to occur*
Aldol condensations take place with certain aldehydes in the presence of dilute alkali. The reaction is:

$$\underset{1 \quad\;\; 2}{RCH_2CHO} + \underset{3 \quad\;\; 4}{RCH_2CHO} \rightarrow \underset{1 \quad\; 2 \qquad\;\; 3 \qquad 4}{RCH_2CH(OH)CH(R)CHO}.$$

The numbers in the above equation show that in order that the reaction can take place carbon atom number 3 must possess at least one hydrogen atom. (This is transferred during the reaction). Methanal has no carbon atom adjacent to the carbonyl carbon atom. 2,2-dimethylpropanal and benzaldehyde have no hydrogen atoms at all on the α atom (the carbon atom adjacent to the CHO group). Ethanal will give the reaction.

13 Answer **B**   *Knowledge of the Cannizzaro reaction*
This involves the simultaneous oxidation and reduction of the aldehyde. Ketones cannot take part in the reaction because they are not easily oxidized. There is nucleophilic attack upon the carbonyl carbon atom by OH^-, followed by the elimination of H^-, which then acts as a nucleophile and attacks another

carbonyl carbon atom:

$$C_6H_5-\underset{\underset{OH^-}{|}}{\overset{\overset{O}{\|}}{C}}-H \;\rightarrow\; C_6H_5-\underset{\underset{OH}{|}}{\overset{\overset{O^-}{\|}}{C}}-H \;\rightarrow\; C_6H_5COOH;$$

$$C_6H_5-\underset{\underset{H^-}{}}{\overset{\overset{O}{\|}}{C}}-H \;\rightarrow\; C_6H_5CH_2O^-.$$

$$C_6H_5COOH + C_6H_5CH_2O^- \;\rightarrow\; C_6H_5COO^- + C_6H_5CH_2OH$$

This reaction is only given by those aldehydes which do not possess a hydrogen atom on the carbon atom adjacent to the CHO group (called an α-hydrogen atom).

14 Answer **D**   *Comprehension of addition reactions of aldehydes*
A sodium hydrogensulphite addition product will generate the original aldehyde or ketone if it is treated with aqueous alkali in low concentration. This method may be used to purify carbonyl compounds. The addition product with propanone would be $(CH_3)_2C(OH)SO_3Na$. The reverse reaction occurs by the removal of a proton from the molecule by the OH^- ion followed by rearrangement and the loss of SO_3^{2-}.

15 Answer **A**   *Comprehension of the cyanohydrin reaction*
The carbonyl compound

$$\underset{R'}{\overset{R}{\diagdown}}\!\!\!\diagup\,C=O,$$

where R and R′ may be alkyl groups or hydrogen atoms,
reacts with HCN to form:

$$\underset{R'\diagup \; \diagdown CN}{\overset{R\diagdown \; \diagup OH}{C}}$$

which upon hydrolysis generates

$$\underset{R'\diagup \; \diagdown COOH}{\overset{R\diagdown \; \diagup OH}{C}}$$

In the case of 2-hydroxypropanoic acid, $R = CH_3$ and $R' = H$. The starting compound is ethanal.

16 Answer **C**   *Comprehension of dehydrogenation*
Dehydrogenation is the removal of hydrogen from a compound. The changes in the formulae that are occurring are:

$C_2H_6O \rightarrow$ **A** $C_2H_4O_2$ **A**   involves gain of oxygen as well as loss of hydrogen.
               B C_2H_4 **B**   involves the loss of water-dehydration.
               C C_2H_4O **C**   involves the simple loss of hydrogen.
               D C_3H_6O **D**   involves the gain of a carbon atom.

Knowledge of the differences in reactions
17 Answer **D**   *between aldehydes and ketones*
Reagents (i), (ii) and (iii) will react only with aldehydes and, therefore, could be used to differentiate between aldehydes and ketones. The restoration of the magenta colour to Schiff's reagent, (iv), is positive for both propanone and ethanal although it occurs very slowly with the ketone. Both compounds, therefore, would give a positive test with reagent (iv).

18 Answer **B**   *Application of knowledge of the formulae of aldehydes*
Aldehyde functional groups can only occur at the *end* of carbon chains. Consequently, there are two sites on the unbranched compound where the CHO group might be. These two sites, however, are equivalent. The compound is hexanal: $CH_3CH_2CH_2CH_2CH_2CHO$.

19 Answer **D**   *Application of knowledge to the formula of aldehydes*
With one branch in the chain, there are many possibilities. These include:
(i) $CH_3CH_2CH_2CH(CH_3)CHO$ (2-methylpentanal).
(ii) $CH_3CH_2CH(CH_3)CH_2CHO$ (3-methylpentanal).
(iii) $(CH_3)_2CHCH_2CH_2CHO$ (4-methylpentanal).

XX Aromatic Compounds

1 Answer **D** *Knowledge of the structure and properties of benzene*
The formula of benzene is C_6H_6 and the structure is cyclic. The six carbon–carbon bonds are all equal in length and the bond angles are also equal. Clearly, therefore, the compound is not a cyclic alkene (cyclohexa-1,3,5-triene), for this compound would have two kinds of carbon–carbon bond, a relatively long single bond and a relatively short double (alkene) bond. The bond angles would not be equal to 120° and the cyclic structure would not be symmetrical as is the case with benzene:

benzene   cyclohexa-1,3,5-triene

The failure of benzene to react rapidly with bromine is due to the absence of double (alkene) bonds. The reaction of benzene with liquid bromine requires the presence of a halogen carrier, such as iron. This gives rise to the complex $[FeBr_4]^-Br^+$ allowing electrophilic attack by Br^+ on the benzene ring. This reaction is classed as substitution, not addition. A similar substitution reaction with chlorine would yield three dichloro compounds:

1,2-dichlorobenzene   1,3-dichlorobenzene   1,4-dichlorobenzene

2 Answer **D** *Knowledge of the structure of benzene*
Bonding in benzene involves carbon sp^2 hybrid orbitals and a carbon p orbital. The sp^2 hybrid results in a planar molecule with bond angles about each carbon atom of 120°. The carbon–carbon bond is formed by the overlap of the sp^2 orbitals and by the π electrons, arising from the p orbitals, which are delocalized over the carbon atoms. The bond is shorter than an aliphatic C—C (alkane) bond, but longer than an aliphatic C=C (alkene) bond.

3 Answer **B** *Knowledge of the electrophilic substitution reactions of benzene*
The aluminium chloride can polarize a chlorine molecule: $Cl^+—Cl^- \ldots AlCl_3$. In an analogous way to the reaction with bromine in question one, a complex ion is formed which allows electrophilic attack by Cl^+ upon the electron-rich ring. The complex ion is $[AlCl_4]^-Cl^+$.

4 Answer **B** *Comprehension of the mechanism of the nitration of benzene*
The electrophile in this reaction is the nitryl cation NO_2^+ formed by the mixing

of concentrated sulphuric and nitric acids:

$$2\,H_2SO_4 + HNO_3 \rightarrow NO_2{}^+ + 2\,HSO_4{}^- + H_3O^+.$$

Attack occurs as follows:

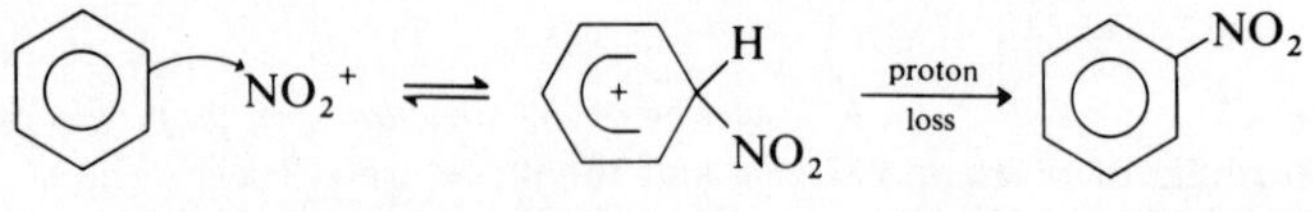

The unstable intermediate involves both the entering group ($NO_2{}^+$) and the leaving group (H^+) and because of the electrophilic attack, the symmetry of the delocalized π structure is disrupted. This shown by the incomplete circle within the ring. The spreading of the positive charge over the molecule helps to stabilize the intermediate compound with respect to the reactants although, of course, the final product, nitrobenzene, is more stable energetically than the intermediate.

5 Answer **B**   *Knowledge of the structure and properties of methylbenzene*
There are four benzene-based isomers of formula C_7H_7Cl. They are:

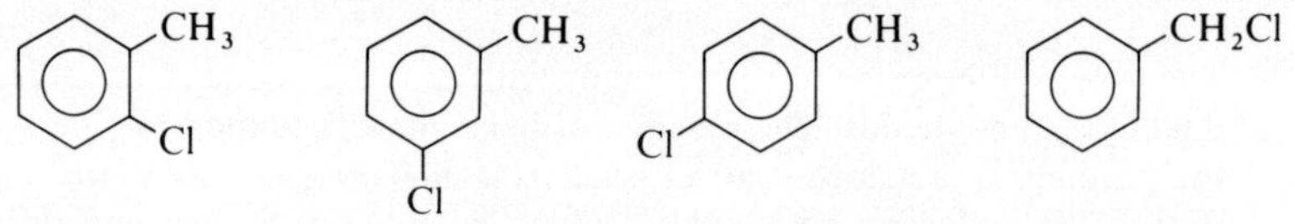

The nitration of methylbenzene is easier than the nitration of benzene because the CH_3 group has a $+I$ inductive effect, thereby activating the ring towards electrophilic attack. The benzene ring is planar, because of the sp^2 hybridization, but the presence of the CH_3 group means that the tetrahedral arrangement of hydrogens about this carbon atom ensures that the molecule is not planar. Benzoic acid can be formed by the treatment of methylbenzene with warm acidified manganate(VII) solution. The purple colour of the manganate(VII) is discharged in this reaction with the formation of $Mn^{2+}(aq)$.

6 Answer **D**   *Comprehension of the formation of diazonium compounds*
A diazonium compound is formed by reacting an aromatic amine with a solution of sodium nitrite ($NaNO_2$) and hydrochloric acid at $0\,°C$. The amine group must be attached directly to the benzene ring. The stability of the diazonium salts (as opposed to similar alkyl compounds, $RN_2{}^+$) lies in the bonding of the nitrogen atom directly to the ring, thus enabling the structure to be stabilized by resonance:

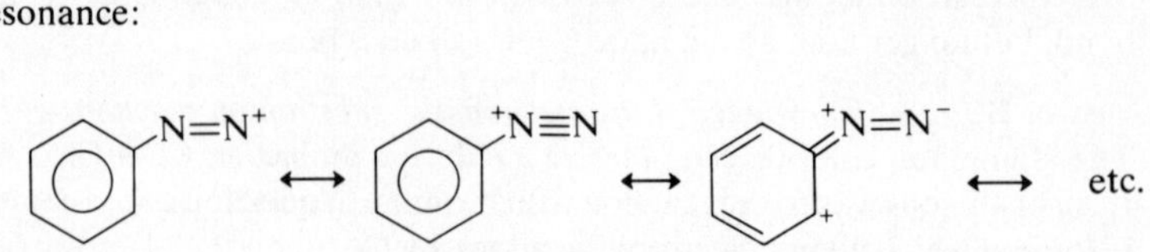

7 Answer **B**   *Application of knowledge of diazonium compounds*
Decomposition of $C_8H_9ClN_2$ produces an unreactive gas and an aromatic compound. This would suggest a diazonium compound. The formula would suggest that it is a substituted benzene derivative. The reaction with potassium

iodide solution backs up this interpretation as iodine can replace the $-N=N^+Cl^-$ structure in a diazonium salt. The decomposition of a diazonium salt in aqueous solution produces a phenol, nitrogen and a hydrogen halide. Response **B** is the only phenol listed. The reaction is:

$$C_8H_9ClN_2 \qquad C_8H_{10}O \quad (Z)$$

8 Answer **B**   *Comprehension of a multi-step synthesis*

9 Answer **D**

10 Answer **C**

The preparation of compound 13 from compound Z is an oxidation, and the reaction of 13 to form Y is a decarboxylation. This suggests the sequence:

$$\text{methylbenzene} \ldots \text{benzoic acid} \ldots \text{benzene.}$$
$$Z \qquad\qquad 13 \qquad\qquad Y$$

The formation of compounds 14 and X is carried out in a nitrating medium suggesting the sequence:

$$\text{benzene} \ldots \text{nitrobenzene} \ldots 1,3\text{-dinitrobenzene.}$$
$$Y \qquad\qquad 14 \qquad\qquad X$$

The production of compound 15 is a reduction. The reagent used will bring about the reduction of one of the nitro groups to form 3-nitrophenylamine. This can then form a diazonium compound, capable of reacting with naphthalen-2-ol to form an azo dye:

$$1,3\text{-dinitrobenzene} \ldots 3\text{-nitrophenylamine} \ldots \text{diazonium salt} \ldots \text{azo dyestuff}$$
$$X \qquad\qquad\qquad 15 \qquad\qquad\qquad W \qquad\qquad\quad V$$

Comprehension of conditions and reagents
11 Answer **B**   *necessary to effect a multi-step synthesis*

12 Answer **E**

13 Answer **C**

14 Answer **A**

15 Answer **C**

16 Answer **A**

Step **E** is the reaction between 2,4-dinitrophenylhydrazine and propanone in which a molecule of water is eliminated during the formation of the hydrazone. Step **D** is a replacement of a bromine atom substituted into the benzene ring by hydrazine and this leads to the second solid product in the sequence. The first solid product is $C_6H_5N_2O_4Br$, which is a bromobenzene with two nitro groups substituted into the ring (step **C**). These will be in positions 2 and 4:

Bromine directs to positions 2 and 4 and one NO_2 group substituted into the molecule at position 2 will also direct the second nitro group into position 4. Step **C**, the nitration is brought about by concentrated nitric and sulphuric acids. The formation of bromobenzene from benzene (step **B**) requires iron to act as a halogen carrier (see question one). Step **A**, the formation of the arene from the alkyne refers, in this case, to the formation of benzene from ethyne by passing it through a red hot tube. This reaction was first carried out in 1870 by Berthelot and leads to other products as well as benzene. The use of the term polymerization here is of dubious accuracy. The term polymerization was originally used to describe the process occurring when a 'monomer' was converted into products which had the same empirical formula but different relative molecular masses, each of which was a multiple of that of the 'monomer'. Carothers (of Nylon fame) in 1931 defined it as '.. intermolecular combinations that are functionally capable of proceeding indefinitely.' Using this latter definition, the production of benzene from ethyne is not polymerization.

17 Answer **B**   *Knowledge of the nitration of arenes*

As we have seen in question four, nitration involves electrophilic attack upon the ring by NO_2^+, the nitryl cation. In a mixed acid medium, the equation is:

$$2\,H_2SO_4 + HNO_3 \rightarrow 2\,HSO_4^- + NO_2^+ + 2\,H_3O^+$$

In concentrated nitric acid (aqueous or organic solvent) the nitryl cation is also thought to be responsible for the nitration:

$$2\,HNO_3 \rightarrow NO_2^+ + H_2O + NO_3^-.$$

Nitrations brought about in dilute nitric acid are thought to involve other nitrating species, e.g. $H_2NO_3^+$.

In mixed acid medium, there are various pieces of evidence supporting the contention that NO_2^+ is the nitrating species.

(i) Salts containing nitryl cations, e.g. $(NO_2^+)(ClO_4^-)$ and $(NO_2^+)[BF_4]^-$ have been shown to be nitrating agents.

(ii) Cryoscopic studies on mixed concentrated sulphuric acid and nitric acid have shown the presence of four ions, satisfying the equation shown at the beginning of this answer.

(iii) Raman spectroscopy has shown an absorption band at $1400\,\text{cm}^{-1}$, which corresponds with that of the nitryl cation.

                                  Knowledge of the factors influencing the ease
18 Answer **D**   *of nitration of arenes*

As nitration involves electrophilic attack upon the ring, the compound possessing the highest electron density on the ring will be easiest to nitrate. Chlorobenzene and nitrobenzene both involve $-I$ groups substituted on to the ring, which tend to withdraw electron density. In addition, the nitro group has a $-M$

effect, which also results in the reduction of electron density on the ring, this time by the mesomeric effect. The chloro group, however, has a $+M$ effect, which tends to increase the electron density, but because the $-I$ effect is stronger than the $+M$ effect, the chloro group acts as a deactivator. With phenol, there is a small $-I$ effect and a larger $+M$ effect with the result that the electron density on the ring is increased with respect to benzene. Phenol can be nitrated using dilute acid.

19 Answer **D**

Knowledge of factors affecting electrophilic attack upon arenes

Bromination involves electrophilic attack by Br^+ and will be facilitated by a large electron density upon the ring. The presence of halogen or nitro groups on the ring deactivates it towards further electrophilic attack (previous question) by reducing the electron density which is present on the ring. Methylbenzene, possesses a CH_3 group which has a small $+I$ effect and essentially no mesomeric effect, and the result is that the ring is activated towards electrophilic attack.

20 Answer **D**

Knowledge of the differences between the OH groups in phenol and ethanol

Phenol cannot be esterified using a carboxylic acid, an acid chloride must be employed. Similarly, bromination cannot be carried out in the simple way that it can with ethanol. Treatment of phenol with bromine water results in the substitution of three atoms of bromine into the ring at positions 2,4 and 6 to form 2,4,6-tribromophenol. The bromine atom cannot replace the OH group in this way. Sodium, however, reacts with both compounds to produce an ionic product involving the $-O^-Na^+$ structure. Whilst ethanol is miscible with water in all proportions at room temperature, phenol is not.

21 Answer **D**   *Application of knowledge to reactions of benzene derivatives*

The oxidation of $C_7H_7NO_2$ to a compound containing two more oxygen atoms and two fewer hydrogen atoms suggests a CH_3 group being oxidized to COOH. Acidified manganate(VII) would bring about this conversion. Tin and concentrated hydrochloric acid is a reducing mixture and could reduce a nitro group (NO_2) to an amine group (NH_2). This fits in with the loss of two oxygen atoms from the molecule and the gain of two hydrogen atoms as specified in the question.

22 Answer **B**   *Knowledge of the reaction of fuming sulphuric acid and benzene*

Fuming sulphuric acid contains a proportion of free sulphur(VI) oxide (SO_3) dissolved in it. This is an electron deficient molecule (electrophile), which seeks out electron density at the benzene ring forming the intermediates

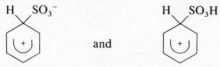

which are in equilibrium. Subsequent rearrangement and proton loss produces

167

23 Answer **A**   *Comprehension of Friedel–Crafts acylation*
The condensation of benzene with an acid chloride or an acid anhydride in the presence of anhydrous aluminium chloride produces an aromatic ketone. The mechanism for the reaction is thought to involve electrophilic attack upon the ring by the cation RCO^+ (in this case CH_3CO^+), which is formed by the interaction of the acid chloride or anhydride and aluminium chloride, e.g.

$$RCOCl + AlCl_3 \rightarrow RCO^+ + [AlCl_4]^-.$$

Responses **B** and **D** would involve the production of esters whilst **C** would merely result in the hydrolysis of the ethanoic anhydride.

                                         Knowledge of ring activation/deactivation
24 Answer **D**   *and directing effects of substituents*
When a second substituent is introduced into a monosubstituted benzene ring, the ease of reaction, compared with that of benzene in forming the original monosubstituted compound, depends on the nature of the substituent already present in the ring. The position of the second substitution is also dependent upon this. The $+I$ effect of the methyl group activates the ring, thus enhancing the ease of electrophilic substitution. The presence of Br or NO_2 in a ring deactivates the ring towards electrophilic substitution ($-I$ effect) but the substituents differ in their effects upon the position of the second substitution. Although NO_2 deactivates the ring because of the mesomeric effect, positions 3 and 5 are deactivated less than positions 2, 4 and 6, and so when further nitration occurs the compound that is formed is 1,3-dinitrobenzene and not the 1,2-or 1,4-compounds. Conversely with bromine, positions 2,4 and 6 are activated relative to positions 3 and 5, although, all positions are deactivated with respect to benzene. With bromine, therefore, the compound formed is 1,2-dibromobenzene.

25 Answer **B**   *Knowledge of the side-chain substitution of methylbenzene*
The side-chain substitution of methylbenzene involves chlorine free radicals formed by the homolytic fission of chlorine molecules. A chain reaction occurs and the products are $C_6H_5CH_2Cl$, $C_6H_5CHCl_2$, $C_6H_5CCl_3$ and HCl. A halogen carrier is only required when substituting chlorine atoms directly into the ring.

26 Answer **B**   *Comprehension of the structure of substituted benzenes*
The isomers are:

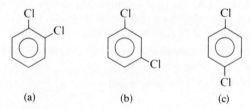

(a)   (b)   (c)

(a) 1,2-dichlorobenzene
(b) 1,3-dichlorobenzene
(c) 1,4-dichlorobenzene

XXI Macromolecules

1 Answer **B**   *Knowledge of the properties and structures of sugars*
Monosaccharides can contain up to six carbon atoms per molecule, e.g. ribose
has five. They do not undergo acidic hydrolysis. Disaccharide molecules contain
twelve carbon atoms per molecule and undergo acidic and enzymic hydrolysis to
yield two monosaccharide molecules.

2 Answer **C**   *Comprehension of monosaccharide nomenclature*
The prefix keto- shows that the molecule contains a ketone group and the prefix
aldo- shows that the molecule contains an aldehyde group. The number of
carbon atoms in the molecule is indicated by the terms tri-, tetr-, pent- and hex-
representing three, four, five and six atoms per molecule respectively.

3 Answer **C**   *Knowledge of the properties and structure of glucose*
Glucose is an aldohexose and shows the typical reaction of an aldehyde with
Fehling's solution. It can be produced, together with fructose from the hydrolysis of sucrose. There are two anomers of glucose. These are two
stereoisomers which have differing configurations about the C_1 carbon atom:

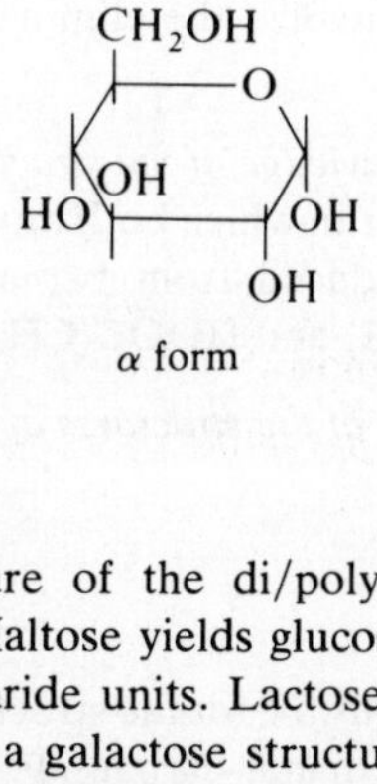

α form β form

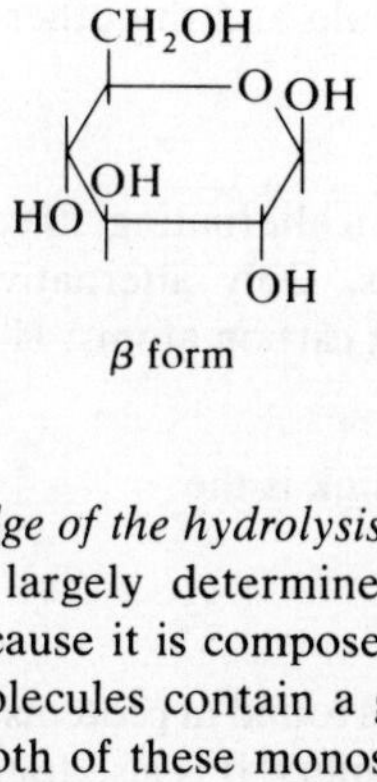

4 Answer **A**   *Knowledge of the hydrolysis of carbohydrates*
The structure of the di/polysaccharide largely determines the hydrolysis
products. Maltose yields glucose only, because it is composed of two glucose
monosaccharide units. Lactose, whose molecules contain a glucose structure
attached to a galactose structure yields both of these monosaccharides upon
hydrolysis with lactase. Hydrolysis of sucrose produces its constituents—glucose
and fructose, whilst the hydrolysis of starch with diastase produces maltose.

                                Comprehension of the structure of a polymer
5 Answer **C**   *in determining its empirical formula*
The empirical formula of a polymer is the repeating unit of which the polymer is
composed.   For   a   simple   polymer   such   as   poly(ethene),
$-CH_2-CH_2-CH_2-CH_2-$, the empirical formula is readily seen to be CH_2.
For the more complex structure, rubber, the basic repeating unit is shown here

between the dotted lines:

$$\text{CH}_2\diagdown\text{C}=\text{C}\diagup\text{H}_3\text{C}\quad\text{H} \qquad \text{H}_3\text{C}\quad\text{CH}_2 \quad \text{C}=\text{C} \quad \text{CH}_2 \quad \text{CH}_2 \quad \text{CH}_2 \quad \text{C}=\text{C} \quad \text{H}_3\text{C}\quad\text{H}$$

The empirical formula is C_5H_8

6 **Answer D**   *Knowledge of polymer arrangements*
A block copolymer consists of regular blocks of alternate monomer molecules. Alternative **C** represents a random copolymer whilst alternative **A** is a linear polymer with only one type of monomer molecule, e.g. poly(ethene). Alternative **B** is a linear alternating copolymer, e.g. nylon 6.6.

7 **Answer C**   *Knowledge of the term 'isotactic polymer'*
When an alkene derivative, $CH_2=CHR$, is polymerized the group R may be present on one side of the polymeric chain (isotactic: iso = same), on alternate sides (syndiotactic) or randomly distributed on both sides (atactic). Of the alternatives, only propene is an asymmetrical alkene ($R = CH_3$). Ethene has the structure $CH_2 = CH_2$, tetrafluoroethene, $CF_2 = CF_2$ and tetrachloroethene, $CCl_2 = CCl_2$.

8 **Answer A**   *Knowledge of free radical polymerization mechanisms*
In a chain reaction, the free radical is conserved (propagated). Only the intermediate shown in response **A** can contribute to the formation of poly(phenylethene). Response **B** involves the decomposition of the phenylethene molecule and the other responses involve the elimination of the free radical.

9 **Answer D**   *Knowledge of the structure of nylon 6.6*
Nylon 6.6 is an alternating, linear copolymer in which both monomers have six carbon atoms. Only alternative **D** is produced from two monomers each possessing six carbon atoms: $H_2N(CH_2)_6NH_2$ and $HOOC(CH_2)_4COOH$.

10 **Answer C**   *Knowledge of the structures of macromolecules*
The peptide link is the

$$-\underset{\underset{O}{\|}}{C}-\underset{\underset{H}{|}}{N}-$$

bond unit. It is found in proteins and is responsible for the structure of the nylon group of polymers. It is also found in the methanal–carbamide resins. Terylene is a polyester and involves repeated ester linkages:

$$-\underset{\underset{O}{\|}}{C}-O-.$$

11 **Answer C**   *Comprehension of the primary structure of proteins*
Amino acids possess at least one carboxylic acid end group and one amino end group. It is these groups which give rise to the peptide link. Representing the amino acids X and Y by $n—X—c$ and $n—Y—c$, where n and c stand for the amino and carboxylic acid end groups respectively, there are two possible

dipeptides that can be formed between X and Y:

$$n-X-Y-c \quad \text{and} \quad c-X-Y-n.$$

When three amino acids are involved, the possibilities become:

$$c-X-Y-Z-n \quad n-X-Y-Z-c$$
$$c-X-Z-Y-n \quad n-X-Z-Y-c$$
$$c-Z-X-Y-n \quad n-Z-X-Y-c.$$

12 Answer D

Comprehension of the structures of molecules required to bring about condensation polymerization

A polymer consists of many monomer units jointed together in chains. Both of the molecules employed in a condensation polymer need reactive end-groups at *each end* of the molecule to ensure that a chain is formed. The formula of the polymer produced is

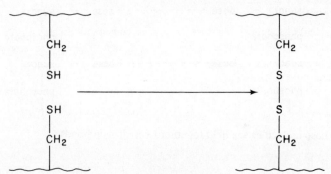

13 Answer A   *Knowledge of cystine links in proteins*

The reaction to form the cystine link can be represented by the equation:

This reaction involves a loss of hydrogen and is classed as an oxidation.

14 Answer B   *Knowledge of condensation polymerization*

Condensation polymerization occurs when two different monomers react to form a polymer chain and a small molecule is also produced. Often this molecule is water and this has given rise to the term 'condensation' polymerisation, although other substances, e.g. hydrogen chloride, may be produced. In this example, the diol can react with a dicarboxylic acid in the following way:

$$n\,HO-CH_2-CH_2-OH + n\,HOOC-(CH_2)m-COOH \rightarrow$$
$$\left[OCH_2-CH_2-O-\underset{\underset{O}{\|}}{C}-(CH_2)m-\underset{\underset{O}{\|}}{C}\right]_n + n\,H_2O$$

This polymer is a polyester. Terylene is an example of a polyester. Other examples of condensation polymers are the nylons and the methanal–carbamide resins.

171

15 Answer **C** *Application of knowledge of the properties of functional groups*

16 Answer **B**

17 Answer **C**

18 Answer **A**

The R_f value in the phenol/water solvent system is a measure of the solubility of the amino acid in phenol. Those amino acids with phenyl or alkyl groups in the molecule will, therefore, have relatively high R_f values whilst the other amino acids will have relatively low ones. The R_f values for the molecules described in the questions are: (15) 0.86; (16) 0.14; (17) 0.84; (18) 0.33.

Those amino acids containing carboxylic acid groups in the side chain will be acidic in that their isoelectric points will occur at low pH values. The isoelectric points for the five compounds are: (15) 5.5; (16) 2.8; (17) 6.0; (18) 5.7.

The isoelectric point of an amino acid is the pH at which electrophoresis cannot take place because of the balance of charges on the molecule.

19 Answer **B** *Knowledge of the structure of DNA*

20 Answer **C**

The following diagram shows part of the linked chain structure of DNA:

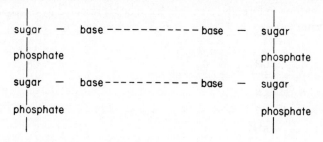

The base molecules are held together by hydrogen bonds

XXII   Analysis

1 Answer **A**   *Comprehension of the reactions of potassium manganate(VII)*
Potassium manganate(VII) acts as an oxidizing agent in acidic solution, being
reduced itself to manganese(II). It will oxidize sulphite ions to sulphate, but will
not oxidize sulphate ions, nitrate ions or iron(III) ions.

2 Answer **D**   *Knowledge of the reactions of dilute nitric acid*
Potassium carbonate and potassium hydrogencarbonate cannot be distinguished by the addition of a dilute acid, as both compounds will produce
carbon dioxide. It is possible, however, to distinguish between the other pairs of
compounds as no reaction will be found with one member of each pair when
dilute nitric acid is added (i.e. the iodide, sulphate and nitrate). The nitrites will
yield traces of brown nitrogen dioxide gas at room temperature. The carbonate
will effervesce with dilute acid.

3 Answer **C**   *Comprehension of the effects of heat and acid on compounds*
Addition of dilute acid to a solid to produce a brown gas suggests that the solid is
a nitrite. Nitrous acid is produced in solution and this decomposes to give $NO(g)$
and $NO_2(g)$. A nitrite is produced by heating a Group I nitrate other than lithium
nitrate. Oxygen is also produced in the reaction.

4 Answer **B**   *Comprehension of the tests for potassium and halide ions*
A yellow precipitate with sodium hexanitrocobaltate(III) indicates the presence
of potassium or ammonium ions. Bromide ions give a cream precipitate of silver
bromide in the presence of silver nitrate solution. Silver chloride and bromide
dissolve in excess concentrated ammonia solution. The bromide will not dissolve
in dilute ammonia solution.

5 Answer **A**   *Application of knowledge to titration results*
The amount of sodium hydroxide in 25 cm³ of the original solution is proportional to 20 cm³ of 0.1 M hydrochloric acid. This neutralization reaction
produces sodium chloride which is equivalent to 20 cm³ of silver nitrate solution
(0.1 M). The sodium chloride in the original solution, therefore, is only
equivalent to 10 cm³ of 0.1 M silver nitrate solution. The molar ratio of sodium
chloride to sodium hydroxide, therefore, is $10:20$ or $1:2$.

6 Answer **B**   *Application of knowledge to experimental data*
Substance P must be barium chloride because it gives two white precipitates.
These are formed with Q and R, one of which must be the sulphate and the other
the carbonate. Q is the carbonate as it effervesces with substance S. Substance S,
therefore, is the acid, leaving R as the sulphate.

7 Answer **D**   *Knowledge of the reactions of carbonyl compounds*
Substance Z contains a carbonyl group because it gives a brightly coloured
product with 2,4-dinitrophenylhydrazine. It cannot be an aldehyde as it does not
react with ammoniacal silver nitrate solution. Substance Z must be the ketone,
$C_2H_5COC_2H_5$.

8 Answer **B**   *Application of knowledge to experimental data*
W and Y are aldehydes as they reduce ammoniacal silver nitrate solution. W is
ethanal because it gives a positive iodoform test (it contains the CH_3CO group).
X is a ketone which contains the CH_3CO group because it gives a positive
iodoform test, reacts with 2,4-dinitrophenylhydrazine and does not reduce
ammoniacal silver nitrate solution. Substance Z does not possess a carbonyl
group, but is oxidized to a compound containing the structure CH_3CO under the
conditions of the iodoform test. This substance is propan-2-ol.

                                  Knowledge of the reactions of manganate(VII)
9 Answer **D**   *ions with organic compounds*
Under the stated conditions, both methylbenzene and propan-1-ol will be
oxidized with the manganate(VII) ions being reduced to manganese(II) ions. An
unsaturated compound such as but-1-ene will decolorize the manganate(VII)
solution in being oxidized to the diol. Propanone cannot be oxidized without
carbon–carbon bond rupture. Acidified manganate(VII) will not do this.

10 Answer **B**   *Comprehension of the differing reactions of aldehydes*
Only ethanal, containing a CH_3CO group, will give a positive iodoform test. The
presence of the phenyl group in benzaldehyde assists the rapid formation of a
crystalline product with sodium hydrogensulphite.

11 Answer **B**   *Application of knowledge to titration results*
The equations show that 2 mole of $CuSO_4 \equiv 1$ mole of $I_2 \equiv 2$ mole $Na_2S_2O_3$. It
follows that 1 mole of $CuSO_4 \equiv 1$ mole of $Na_2S_2O_3$. 25 cm^3 of 0.1 M sodium
thiosulphate(VI) solution is required if 25 cm^3 of 0.1 M copper(II) sulphate is
used.

                                  Knowledge of the determination of secondary
12 Answer **A**   *and tertiary protein structure*
The secondary structure of a protein is the way in which the amino acid residues
in the molecular chain are arranged with respect to each other. The chain is
usually either folded, coiled or puckered and many proteins have their chains of
amino acids coiled into a helix—the α helix. The tertiary structure of a protein is
the way in which the coiled or folded chains are arranged in space to form a
three-dimensional molecule. Both these structures are determined by X-ray
diffraction.

13 Answer **A**   *Comprehension of a mass spectrum*
Mass spectrometers generate positively charged ions whose path through the
machine is determined by their mass: charge ratios and the applied electric and
magnetic fields. The mass: charge ratios for the ions A to D are: **A** 15:1; **B**
14:2; **C** 16:1; **D** 17:1.

14 Answer **A**   *Comprehension of titration results*
From the equation it is possible to see that 0.01 mole of ethanedioate ions
require 0.004 mole of manganate(VII) ions for complete oxidation to occur.
This is equivalennt to 40 cm^3 of a 0.1 M solution. The other 20 cm^3 of
manganate(VII) ions is used in oxidizing the iron(II) to iron(III).

15 Answer **C**   *Knowledge of protein analysis*
The dinitrophenylamino acid is formed by the hydrolysis of the peptide bond.

This is brought about by refluxing with 6 M hydrochloric acid. Aqueous alkali has no effect here. The result of using concentrated hydrochloric acid is the hydrolysis of much of the molecule. The bond formed between the dinitrophenyl group and the amino acid chain is stable to hydrolysis under these conditions.

16 Answer **C**   *Knowledge of solubilities of salts*

Unlike the other silver halides, silver fluoride is water soluble. Alkali metal sulphates, halides and ethanedioates are also soluble. Barium sulphate, however, is insoluble and forms as a white precipitate.

XXIII Practical Techniques

1 Answer **A**

Knowledge of the conditions necessary in order to test for halide ions

If ions other than halides are present in a solution, e.g. sulphite, carbonate and hydroxide, they will form precipitates with silver ions. The nitric acid is added to react with these ions so that they will not form a precipitate with the silver nitrate solution.

2 Answer **D**

Knowledge of the conditions necessary for the oxidizing reactions of manganate(VII) ions

Dilute sulphuric acid is usually used to provide the acidic medium required by the manganate(VII) ions in their oxidizing reactions. Any alternative must not be oxidized by the manganate(VII) ions during the reaction, e.g. $H_2S(aq)$ and $HCl(aq)$ would be oxidized. The acid itself must not be an oxidizing agent, e.g. $HNO_3(aq)$.

3 Answer **C**   *Knowledge of steam distillation*

The technique is useful for separating water and a liquid such as phenylamine which is immiscible with water. The organic liquid must not have a boiling point which is close to that of water. Boiling occurs when the sum of the vapour pressures of the components of the mixture equals the external pressure. This occurs below 100 °C at 1 atm pressure.

4 Answer **C**   *Comprehension of drying techniques for organic materials*

Because of the acidic nature of benzoic acid, it cannot be dried with a basic material such as calcium oxide or sodium hydroxide; nor can it be dried with a carbonate with which it could react.

5 Answer **D**

Application of knowledge to the interpretation of titration results

The ratio, volume of $AgNO_3(aq)$: mass of Z can be used to check the consistency of the titration results. The values are: **A** 150; **B** 150; **C** 150; **D** 147.5.

6 Answer **B**   *Comprehension of the use of indicators in acid/base titrations*

The end point of a titration between a strong base and a weak acid occurs at a pH greater than 7. In order that an indicator can accurately reflect the end point, the pH range of the indicator must cover the pH range at which the end point occurs. In this example, phenolphthalein, with a pH range of 8.2 to 10.0 is suitable.

7 Answer **C**   *Knowledge of the preparation of phenylamine*

The tin(IV) chloride, formed in the preparation of phenylamine, combines with chloride ions to form the complex ion $[SnCl_6]^{2-}$. This then combines with the phenylammonium ions, $C_6H_5NH_3^+$, which are present in the acidic medium of the reaction to form $(C_6H_5NH_3^+)_2 \cdot [SnCl_6]^{2-}$. The addition of sodium hydroxide liberates the phenylamine from the complex.

8 Answer **D**   *Comprehension of the drying techniques for organic materials*

Phenylamine is a basic substance and, therefore, cannot be dried with an acid or a substance which can be converted into an acid by hydrolysis ($AlCl_3$ or P_4O_6). The basic calcium oxide is the only possible drying agent for phenylamine shown here.

9 Answer **C**   *Application of knowledge to experimental information*
If 100% conversion takes place, 0.1 mole of acid chloride will produce 0.1 mole of ester. However, 6.6 g of ester is only 6.6/88 mole = 0.075 mole. The percentage conversion, therefore, is $0.075/0.1 \times 100 = 75$.

Knowledge of the techniques of preparation
10 Answer **B**   *of acid chlorides*
The reaction of sulphur dichloride oxide ($SOCl_2$) with propanoic acid yields sulphur dioxide and hydrogen chloride as by-products. These gases are easily removed from the acid chloride formed. The use of phosphorus pentachloride produces phosphorus trichloride oxide (phosphorus oxychloride) which boils at 107 °C.

Comprehension of the techniques necessary to
11 Answer **D**   *use a melting point apparatus*
A wet or impure sample will result in a lowering of the melting point of the sample. A slow heating of the oil should result in an accurate measurement of the melting point. However, if the oil is heated too quickly, the thermometer readings can be changing so rapidly that, by the time that the solid has melted, the temperature indicated by the thermometer is higher than that at which the solid actually melted. The oil should be stirred during the measurement.

12 Answer **C**   *Comprehension of percentage yields*
The expected yields are:

Step (i)   0.4 mole ($1 \times \frac{40}{100}$).

Step (ii) 0.16 mole ($0.4 \times \frac{40}{100}$).

Step (iii) 0.064 mole ($0.16 \times \frac{40}{100}$).